100 birds to see in your lifetime, today!

WATCHING WATERBIRDS

with **Kate Humble** & **Martin McGill**

Edited by Malcolm Tait

BLOOMSBURY WILDLIFE
LONDON · OXFORD · NEW YORK · NEW DELHI · SYDNEY

BLOOMSBURY WILDLIFE
Bloomsbury Publishing Plc
50 Bedford Square, London, WC1B 3DP, UK

BLOOMSBURY, BLOOMSBURY WILDLIFE and the Diana logo are trademarks of
Bloomsbury Publishing Plc

First published in 2011 by A&C Black, an imprint of Bloomsbury Publishing Plc

A catalogue record for this book is available from the British Library

ISBN: 978-1-4729-6703-9

2 4 6 8 10 9 7 5 3 1

Design by Lou Milward
Printed and bound in China by RR Donnelley

To find out more about our authors and books visit www.bloomsbury.com
and sign up for our newsletters

'If it looks like a duck, and
quacks like a duck... we
have at least to consider
the possibility that we have
a small aquatic bird of the
family Anatidae on our hands'

Douglas Adams

Contents

Welcome to the wetlands

Wetlands are curious places. At first glance, it can be difficult to find anything appealing about them at all. Flat, bleak, few trees, damp, featureless and all too often experienced under leaden skies and being whipped by a howling gale. No surprise then that they also appear, at first glance, to be devoid of many, if any, living things at all.

But look a bit closer and wetlands are as bustling and busy as any city centre on a Saturday. There are birds everywhere; big ones, little ones, waddling ones, diving ones, birds that soar, birds that furiously peck the ground like manic sewing machines, and some that just hang about in groups doing not much, like bored teenagers. And once you realise just how much is going on out there it is all too easy to become hooked. You can spy for hours on the comings and goings of all these birds, the scraps, the flirting, the feeding, the fighting – and then something else happens. You realise that you want to know what they all are and what they're up to out there.

Well, you do if you're me, and I know I'm not alone. And I also know I'm not the only one who finds wetland birds utterly confusing when it comes to identifying them, and field guides – excellent when you have a good basic knowledge – are frankly overwhelming when you are starting from scratch. But I'm lucky. I know Martin.

Martin McGill works at Slimbridge in Gloucestershire, the headquarters of the Wildfowl & Wetlands Trust (WWT) which was started back in 1946 by the legendary Peter Scott. I met Martin in 2007 when we were both ringing Whooper Swans – and when I saw what looked like a rather pretty duck, he saw a Wigeon, probably an American one (very exciting, shouldn't have been there) because it had two extra grey feathers under its right eye, or something.

Anyway, it became patently obvious that this man knows his birds and has the patience of the most patient of saints. So I asked him if he would help me write a book for, well, me, if I'm honest, but also for anyone else who wants to be able to sit in a hide and know they are looking at a Whimbrel, not a Curlew, or a Ruff, not a Redshank. Then, satisfied they know what they are watching, they can simply get on with enjoying the spectacle. With a large flask of tea. It can get very chilly in those hides after an hour or two…

Kate Humble

When Kate proposed this book I was interested immediately. Yes, there are many field guides which cover everything, and some of them are brilliant – but sometimes they provide a bit too much to take in for beginners. I know first-hand how visitors who are new to birding can get put off by the vast array of birds in front of them and think they will never get to grips with them all. By covering what are essentially wetland birds along with other species that can be seen at wetland areas, we hope to open up some of the mystery and make watching birds a little more accessible to all beginners.

This project has been an eye opener for me as I find it all very easy after 30 years of birding, but the combination of my experience gained in the field, with Kate being fairly new to it all, works well. The essential ingredient is enthusiasm of which both Kate and I have plenty. Anyone who wants to learn more about anything needs a good helping of enthusiasm.

Birds are fantastic, and wetlands are the best place to watch them. I am fortunate enough to be outside most of the year and see wetlands through the seasons; the winter spectacles of huge flocks, spring and autumn migration and the breeding season. All weather, all times of the day. There is never a dull time, always something to see, focus on, get the most out of and enjoy.

So, let's get to it.

Martin McGill

Great Waterbird Challenge

Birdwatching comes in various forms and guises. For some, the pleasure of watching Blue Tits and Greenfinches plucking peanuts from a feeder is as far as they want to go. And fair enough. Garden birds are almost always on hand, and can be enjoyed from the comforts of one's own home. It's a great way of passing a few hours, and looking after your local wildlife into the bargain.

But the very fact you're reading this book suggests that you're looking to take the next step, and good for you. Birdwatching, once it's in the blood, stays there. It draws you out, looking for new experiences each time you raise your binoculars.

The more you learn, the more you want to learn, and this is why Martin and I have set up the Great Waterbird Challenge (GWC). There are two ways in which you can play the game, and each is designed to help you get a feel for birdwatching, to increase your powers of identification and, ultimately, to get even more from this most wonderful of hobbies.

GWC 1 As you leaf through the pages of this book, you'll notice that each of the key birds included has a GWC score. Martin and I have given points to each bird based on the likelihood of your seeing it. The marks are out of five, and you score one point for a common bird, five points for a rare bird. We've included a checklist on page 252 so you can keep your tally as you go. Set yourself a target – say, **50 points the first time you try wetland birdwatching**, and keep trying to improve that total as you go. Once you hit **100 points**, then you know you're getting good. And it'll also be time to move to step 2 of the GWC.

GWC 2 This is the big one. Once you've scored 100 points in a day, you'll be ready for the ultimate challenge. You'll have honed your skills, based on waterbird watching, and can set yourself the target of achieving, not 100 points, but **100 species of any type of wild bird seen in a single day!**

Martin and I set out one autumn day with just such a target in mind – and you'll be able to follow our progress as you read through this book.

And whatever you do – enjoy yourself! You're on the brink of an enduring hobby that will last you a lifetime.

How to use this book

GWC score **3**

nacle Goose

GWC score Your points scored by sighting this bird. The commonest score one point, the rarest, five

Wingspan This chart measures each bird's wingspan, and also shows you how that span compares to that of the familiar Mallard

me: *Branta leucopsis*

e size: 64cm

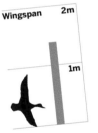

Wingspan   2m

1m

me resident Barnacle geese here in Britain, but far
the Canadas. Most of them migrate here from the
tober and stay around until April. They fly in long lines,
apping calls can be heard at some distance. WWT's
ck is a good place to see them in large numbers, and the
th and Isle of Islay are among their main strongholds
winter months.

smaller, less rangy birds than Canada Geese, and look,
nce, much more black and white, whereas the Canada
ppears brown.

than a chinstrap, the Barnacle Goose has a black neck,
in a very definite white line at the breast and the face is also
white.

Where and when?

Seasonal distribution:
Green: Resident; where the bird may be seen throughout the year and usually breeds.
Yellow: Summer visitor; where the bird may be seen in summer and usually breeds.
Blue: Winter visitor; where the bird spends the winter, but does not breed.
Red: passage migrant; where the bird visits during spring and autumn migration.
The numbers refer to the WWT centres opposite.

Where and when?

MARTIN'S QUICK QUIZ

The Barnacle Goose gets its name from a fascinating, but ancient theory. Can you guess hat it is? Answer on page 251.

Which WWT centre is which?

1. CASTLE ESPIE
Meet the largest collection of wildfowl in Ireland, then complete your day by visiting the shop and Loughshore café.
028 9187 4146
info.castleespie@wwt. org.uk

2. CAERLAVEROCK
From dawn to dusk, January to December, in fair weather and in foul, its open, coastal landscape and wide skies are full of the sights and sounds of nature.
01387 770200
info.caerlaverock@wwt. org.uk

3. WASHINGTON
A diverse reserve of wetland, woodland and meadow – home to endangered wildfowl, waders, flamingos, cranes, frogs, dragonflies, bats and even goats.
0191 416 5454
info.washington@wwt. org.uk

4. MARTIN MERE
Watch internationally important numbers of ducks, geese and swans gathering in winter to form spectacular feeding flocks.
01704 895181
info.martinmere@wwt. org.uk

5. WELNEY
Wildlife spectacles abound here, particularly in winter. And with its eco-friendly visitor centre, Welney is a flagship in sustainable living.
01353 860711
info.welney@wwt.org.uk

6. NATIONAL WETLAND CENTRE WALES
Stretching over 240 hectares, this magnificent mosaic of lakes, pools and lagoons is home to wild species as diverse as dragonflies and egrets.
01554 741087
info.llanelli@wwt.org.uk

7. SLIMBRIDGE
Opened in 1946, Slimbridge is the headquarters of WWT. Its proximity to the Severn Estuary makes it one of the country's great wildlife havens.
01453 891900
enquiries@wwt.org.uk

8. LONDON WETLAND CENTRE
This international award-winning attraction, situated in Barnes, is the best urban site in all of Europe to watch wetland wildlife.
020 8409 4400
info.london@wwt.org.uk

9. ARUNDEL
Nestling in the heart of the Arun valley; with the dramatic backdrop of Arundel Castle, these 65 acres of spectacular reserve are home to rare and beautiful wildlife.
01903 883355
info.arundel@wwt.org.uk

What did you say?
A quick guide to some of the main areas of a bird's plumage and bill, as used in identification.

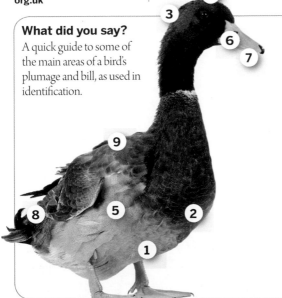

1 Belly Area below the breast to just behind the legs

2 Breast Area below the throat running down to the belly

3 Crest A tuft of feathers on the head of some species

4 Crown The area behind the forehead

5 Flank The side of the body

6 Gape The inside part of the bill

7 Mandible (upper and lower) The upper and lower halves of the bill

8 Speculum A patch of brightly coloured feathers on the wings of most dabbling ducks

9 Upperparts The bird's back

Great Waterbird Challenge
Part 1: An early start

'**S**o I'll see you at 7am,' said Martin. 'And we'll have to start immediately. No time for coffee or anything.'

'Of course, of course. I'll bring a thermos. Drink it as we go. I'll be there. Fresh as a daisy.' And I was. Right on the nose of 7am, I rolled up clutching thermos, notebook and binoculars. 'We've made a good start,' I called as I walked up to Martin. 'You ready?' Of course Martin was ready. This was our big day, our very own Great Waterbird Challenge. We'd set a target of 100 bird species to see across all parts of the reserve at WWT's Slimbridge headquarters and, as I rapidly discovered, he was leaving as little as possible to chance. 'Right then,' he said, rubbing his hands. 'I thought we'd start over at the duck hides. Kick off with the waterfowl. Get some quick ticks under our belts.'

Within a few minutes we were settled on a bench peering through the narrow windows of a hide, as the morning sun bounced off the water in front of us. He was right about the quick ticks. Within what felt like just a few moments, we had 10 ticks in our books. There were **Mallards** (p46) of course, and the familiar shape and pattern of that well known duckpond duck, with its quirky kiss-curl of a tail, was dotted all over the water. We picked out a handful of **Teal** (p49), delicate little creatures that hugged the bank, and several **Shovelers** (p55), too. I rather liked the Shoveler, its distinctive combination of rich burgundy breast and bizarrely spatulate bill giving it an otherworldly appearance – at least in the company of its dabbling peers.

Suddenly, out of the corner of my eye I noticed something surface. A **Tufted Duck** (p68) had emerged from under the water. 'One of the diving duck species,' whispered Martin. 'So's the **Pochard** (p70) – you can see a couple of them towards the back.'

With **Mute Swans** (p16) sailing gracefully and confidently near a reedbed, **Canada Geese** (p34) and **Greylag Geese** (p26) strutting in a field a couple of hundred yards away, and **Wigeon** (p52) in the foreground, grazing on the grass only a few feet from the hide, I got a sudden impression of how Sir Peter Scott must have felt all those years ago when he first came to Slimbridge just after the war. He immediately realised it had the potential to be a true paradise for ducks, geese and swans. Sitting silently, watching them as they go about their business, each feeding in a different way, each occasionally honking or squawking or grunting or even whistling, you really do feel the years fall away. No wonder we all love feeding the ducks on the ponds of parks and villages – they have a timeless quality about them, and give you a sense of peace.

I was jolted from my reverie by a nudge from Martin. 'See that flock of **Gadwall** (p58) over there?' he said. 'Just behind them is a **Pintail** (p60), probably only recently arrived for the winter season, I'd imagine.' Although they're not particularly unusual birds, it was our first unexpected tick. You're almost guaranteed birds like Mallards and Mute Swans, but a Pintail is one of those 'right place at the right time' fellows.

We had 15 ticks under our belt in just a matter of minutes, and I muttered something about being through by lunchtime. Martin glanced at me. 'Don't count your chickens,' he grinned. Heh heh. Always good to go birdwatching with someone who has a sense of humour.

Now turn the page to find out more about the ducks, geese and swans – the waddling birds.

HUGE WADDLING BIRDS

I'm going to start with swans, because almost all of us will have seen a swan at some time or another. We all know they are big and white, and look elegant and graceful on water, although a bit ungainly and bottom-heavy on land. There are three types of swan that are commonly seen in Britain, although the Mute Swan is the only one you'll be able to see all year round. The other two – the Whooper and Bewick's Swans – are often known as wild swans, and they only visit us during the winter months.

Mute Swan – Bill more orangey than other two swans (p16)

Whooper Swan — Yellow on bill projects further than on Bewick's (p20)

Bewick's Swan — The baby of the trio, in terms of size (p22)

Mute Swan

Scientific name: *Cygnus olor*

Approximate size: 160cm

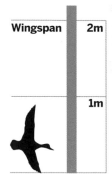

Wingspan 2m

1m

Mute Swans can be seen all over the country and almost anywhere where there is water and food – canals, rivers, lakes, gravel pits. If you see a swan on a pond in an urban park, it will, almost certainly, be a Mute Swan. They are our biggest British bird and they are, as their name suggests, mute. Well, almost. They will hiss at you, neck extended, when they are cross, and they might grunt a bit at each other from time to time, but the noisiest thing about them is actually their wings. You can hear a Mute Swan as it flies overhead because the feathers at the end of its wings make a wonderful 'whooshing' sound as they move through the air. There is a story that Mute Swans sing beautiful, haunting songs when they are in the throes of dying. I can tell you: it's rubbish.

A Mute Swan lives for about 15 years and mates for life. If its mate dies it might, as Martin puts it, mope about looking gloomy for a bit, but then it will find another mate so it can get on with the all-important task of breeding. The female (known as the 'pen', the male is a 'cob') will lay up to six eggs and the cygnets will hatch out greyish brown and won't take on their full adult plumage and bill colouring until they are around two years old. Mute Swans have three unique quirks. When they are defending a territory or showing off to a mate, they do something called 'busking'.

Where and when?

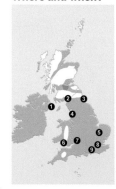

This doesn't involve a guitar and standing on a street corner; in the case of swans it describes the action of arching the wings over the back, spreading out the feathers, so the swan looks even bigger and more regal. Rather like a dog putting its hackles up.

Mute Swans show little, if any, fear of humans and will often go on the attack rather than fleeing from people. We are all told when we are kids that a swan can break your arm with one beat of its wing. Martin clearly didn't listen to that cautionary tale because a swan once broke his finger!

When driving away intruders from its territory, a Mute Swan will push back with both feet simultaneously, rather than paddling them alternately

MARTIN'S QUICK QUIZ

If you can see a pair of Mute Swans, then one will probably have a larger knob on its bill than the other. Which do you think it is, the male or the female?
Answer on page 251.

Whooper Swan

Scientific name: *Cygnus cygnus*

Approximate size: 152cm

Wingspan	2m
	1m

Whooper Swans migrate to Britain from their breeding grounds in Iceland. They come here for a winter holiday and respite from the really nasty weather they get further north, rather like we might go somewhere warm for Christmas. They start arriving during October and they leave to go back to Iceland in late March and April, so if you see a large white swan-looking bird in July it will not be a Whooper Swan.

Their bills are yellow and black, not orange and black, and they don't grow that knobbly, fleshy lump at the base of the bill like the Mutes. The bill patterns are unique to each bird, so when you get to know a group of them really well, you won't just know it's a Whooper, but also that it is 'Fred' (or whatever you've named it).

Whoopers are not quite as big as Mute Swans and they don't have that rather arrogant, imperious air about them. I think they look a bit cheeky and mischievous, as if they are constantly thinking up practical jokes. They migrate as families, non-stop, 800 miles from southern Iceland, and when they arrive they put on a wonderful display of honking, neck stretching and wing flapping to each other known as 'pair bonding' – like giving a member of your family a hug and a kiss when you get to Arrivals at an airport.

Where and when?

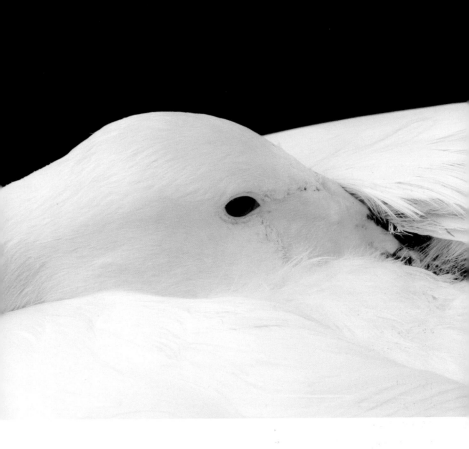

January is the best time of year to see Whoopers

Like Mute swans, Whoopers mate for life, have a slightly longer lifespan of 15-20 years and will hatch about three to five cygnets each year. The cygnets, which migrate with the adults, will be a bit grey and grubby-looking without the bright yellow and black bills, and won't get their full adult plumage until they are two years old.

During the day they will often fly away from the water to graze out on grassland, but Whoopers are particularly fond of root crop fields. Martin says if you see a field of spuds with a lot of happy big white birds in it, they'll probably be Whooper Swans.

Bewick's Swan

Scientific name: *Cygnus columbianus*

Approximate size: 120cm

Wingspan 2m

1m

Now on to the third of our swans, the smallest and arguably the cutest looking. I know, I know. It does look frighteningly similar to a Whooper Swan. But don't panic. There are enough differences to make it possible to tell them apart and one of them is really simple.

Personally, I think the Bewick's has a sort of 'butter-wouldn't-melt, aren't-I-adorable?' look about it. A different sort of expression altogether from the cheeky Whooper, but I concede that could be very difficult to establish at a distance, so let's look at those bills again. Both are black and yellow, but the Bewick's has no yellow below the nostril line. Ever. They, too, have unique bill patterns, but unlike the Whoopers, the bill is always black below the nostril. The best way to remember which is which? The yellow stretches further, just like the 'oo' of Whooper.

Where and when?

It's easier than you'd think to tell the Whooper (above) from the Bewick's (left)

Bewick's Swans migrate too, arriving around October or November and leaving around mid-March. They come here from Russia, where they breed. Martin tells me they tend to have slightly smaller families than the Whoopers – two to three cygnets, and their lifespan is the same at 15-20 years, although the oldest known Bewick's Swan is a positively geriatric 28!

They, too, will honk at each other, and display to establish pair bonds and territory. They graze out in fields, but they will also feed in water, sometimes paddling their feet up and down in the water to get vegetation to rise from the bottom.

BIG WADDLING BIRDS

If it's smaller than a swan, and bigger than a duck, it's a goose. Well, that's easy. And the next stage is easy, too. There are two main types of wild goose found in the UK, and they fall into two colour camps. If they're predominantly grey or grey-brown, then you're in the right section now. If they've got clear areas of black on them, then turn to page 33.

We're starting, though, with the grey geese, four in all. Beaks, legs and feet will be key points of identification for you, so take note of the shades of pink and orange on display.

Greylag Goose — Big with a carrot on its face! (p26)

Bean Goose — Dark head and body, orange bill band (p32)

Pink-footed Goose
— Smaller than the others with dark head and bill (p28)

White-fronted Goose
— That white patch behind the beak is the giveaway (p30)

Greylag Goose

Scientific name: *Anser anser*

Approximate size: 82cm

Wingspan   2m

1m

Greylags are what I think of as proper geese, like the ones you see in farmyards, or in children's books. There is a very good reason for that. Those white, waddling, farmyard geese with their bright orange bills are all descended from Greylags. And it is that beak which helps you tell a Greylag apart from the other grey geese. Martin says they look like they've got a carrot stuck on the front of their faces. Not a very poetic description, but one that you don't forget! So, quite simply, if the goose you're looking at is predominantly greyish brown with a very prominent orange beak, it's a Greylag.

Some Greylags migrate to Britain from Iceland and the Continent, but others are here all year round. This is great news because as recently as 100 years ago, they were extinct as a breeding species in the UK, but have since re-established. There is an Eastern race that is larger, paler and has a pink bill, but that tends to be very rarely seen in the UK.

The Greylag is the largest wild grey goose in all of Europe

Where and when?

Pink-footed Goose

Scientific name: *Anser brachyrhynchus*

Approximate size: 68cm

Wingspan **2m**

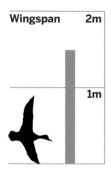

1m

Sadly, you can't simply identify a Pink-footed Goose by looking at its feet and saying 'they are pink, therefore it is a Pink-footed goose', because Greylags have pink feet, too.

So how else do you tell a Pink-foot from a Greylag? A Pink-footed Goose is smaller, and generally more brown than grey.

It has a short neck and a small head, and I think this makes its body look as if it's too large for the rest of it. The head is dark, almost chocolately brown, and the bill is small and dark with pink to pinkish-orange patches.

It grazes in a different way from a Greylag, picking delicately rather than just ripping at the grass. The Pink-feet migrate to Britain from Greenland and Iceland, arriving in September and leaving around the month of May.

Where and when?

You will see Pink-feet in their greatest numbers in the UK in October and November

White-fronted Goose

Scientific name: *Anser albifrons*

Approximate size: 72cm

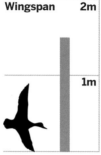

Wingspan   **2m**

1m

Now, if you were to identify a goose simply by its name, you would imagine that this goose has a bright white breast, right? Well, it doesn't. The white front refers to a white splodge on the face, which can be easily missed unless the goose is looking at you head on. In the UK we have two races of White-fronts, the European with its pink bill, and the Greenland with a longer, orange bill, darker plumage and on average heavier belly bars.

It doesn't have as heavy, prominent bill as the Greylag and it is more brown than grey, but the thing that really helps tell these geese apart is the black stripes across the belly. These are particularly noticeable on dominant male birds, but all adults have them.

These geese breed in Russia and Greenland and come to Britain in December, leaving in March. The best time to see the European race is in January, but fewer are being seen in Britain, because warmer winters in places like the Netherlands and southeastern Europe mean they are often choosing to spend their winters there. Greenland birds are found in Ireland, the west coast of Scotland and small families or groups can turn up anywhere, particularly in November.

Where and when?

KATE'S TOP TIPS

With those black stripes across the belly, I think it looks like they've sat on a hot griddle

Juveniles lack the belly stripes

Bean Goose

Scientific name: *Anser fabalis*

Approximate size: 75cm

Our final grey goose is the wonderfully named Bean Goose. We have two races that are regularly recorded in Britain. They, too, come to spend the winter here from Russia and Scandinavia, but only very few – on average about 30-50 Tundra Bean Geese and two small wintering flocks numbering 200-300 Taiga Bean Geese, so it is very exciting if you do spot one.

They do look quite similar to White-fronted Geese, but most don't have the white splodge above their beaks, or the black belly bars. They have a dark head and graze more like a Greylag – tearing at the grass, whereas White-fronted Geese are pickers.

Wingspan 2m

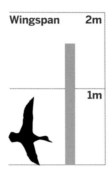

1m

Where and when?

Not all geese are grey or greyish-brown, like the birds we've looked at on the previous pages. In fact, if you've only ever seen one wild goose in your life, there's a good chance it was one of the ones you'll find in this section. It's now time to say hello to the black-and-whites, and I'd imagine that the first of them, the Canada Goose, is already pretty familiar to you.

There are three main species of British goose that have strong black markings on them, and as you'll soon find, they vary quite significantly in size.

anada Goose — Huge oose found all over he country (p34)

Barnacle Goose — Compact version of the Canada, mainly here only in winter (p36)

Brent Goose — Plenty of black, and no white on the face (p38)

Canada Goose

Scientific name: *Branta canadensis*

Approximate size: 95cm

The naturalised race is the biggest of all the geese and is seen all year round, not just in wetland areas, but frequently in urban parks, and grazing out on playing fields. They were released in Britain as an ornamental species and are quite common.

The body is mainly brown with a pale breast. It has a very long neck and bill, the neck is black and it has a white chinstrap. The goose it is most easily confused with is the Barnacle Goose, so let's compare. How to remember? C-anada Goose has the C-hinstrap.

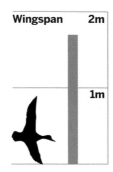

Wingspan   2m

1m

There are many races of this bird and some forms are very small, even smaller than the Barnacle! These smaller races can migrate from Arctic North America with other geese and do sometimes turn up among Barnacle Geese and Pink-footed Geese at WWT Caerlaverock and Martin Mere. They may be smaller than our standard Canada Goose, but their markings are very similar.

Where and when?

Barnacle Goose

Scientific name: *Branta leucopsis*

Approximate size: 64cm

Wingspan   2m

1m

There are some resident Barnacle Geese here in Britain, but far fewer than the Canadas. Most of them migrate here from the Arctic in October and stay around until April. They fly in long lines, and their yapping calls can be heard at some distance. WWT's Caerlaverock is a good place to see them in large numbers, and the Solway Firth and Isle of Islay are among their main strongholds during the winter months.

They are smaller, less rangy birds than Canada Geese, and look, at a distance, much more black and white, whereas the Canada Goose appears brown.

Rather than a chinstrap, the Barnacle Goose has a black neck, ending abruptly at the breast and the face is also mainly white. Look out, too, for a whitish belly and flanks.

Where and when?

MARTIN'S QUICK QUIZ

The Barnacle Goose gets its name from a fascinating, but ancient theory. Can you guess what it is? Answer on page 251.

KATE'S TOP TIPS

I think the Barnacle Goose looks as if it is wearing a black scarf

Brent Goose

Scientific name: *Branta bernicla*

Approximate size: 58cm

The other goose that falls into the black goose category is probably the most complicated of the group. The Brent Goose is divided into a number of very distinct races, but the good news is that their names are the best of clues in telling them apart.

If I were to ask you what the difference was between the Pale-bellied Brent Goose and the Dark-bellied Brent Goose, I'm willing to bet you'd have a pretty good idea, even if you'd never seen the birds. And you'd be right.

The Dark-bellied Brent has a black head and neck, merging into a dark-grey body, with a white patch and a smudgy grey path on its flank. The Pale-bellied Brent has a very distinct line between neck and breast, the body is a paler grey and the patch on the flank is more distinct and whiter. These are the smallest of our geese, and they graze out in the fields.

Pale-bellied Brent Goose

Wingspan   2m

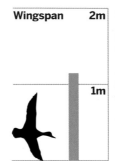

Where and when?

Dark-bellied
Brent Goose

Brent Geese are most commonly seen near the coast where they eat seaweed and eelgrass

Geography can help. The Dark-bellied migrates from Russia, arriving in October, leaving in March, and it tends to spend its time around the east and south coast. The Pale-bellied migrates from Arctic Canada, arriving in October, leaving again in April, and is most commonly seen in Ireland and western Scotland. So, even if you're not quite sure just how pale that belly is that you're looking at, the coastline you're standing on should help you work out which race of Brent you're looking at. Check flocks carefully, however, as small numbers of each race can mingle with others.

Not seen it yet? Perhaps you've seen a goose that doesn't quite look like any we've covered so far? There's a good chance it'll be on these pages that Martin has put together, based on his sightings at WWT centres.

In Britain, we have a number of species that occur which are not actually native, which means they were brought here or escaped and established themselves and breed freely. These self-sustaining populations are now on the British list. Most of them are exotic-looking and easy to identify.

Lesser and Greater Snow Goose

This vagrant goose has a white and blue phase; when it turns up in the UK, it tends to be among other goose flocks. It is most likely to be seen with Pink-footed or Greenland White-fronted Geese, but does also occur among Greylag Geese. So the best place to see one would be at Martin Mere, although they have been recorded at many of WWT's centres.

Lesser White-fronted Goose

This beautiful little goose is, or was, something of a Slimbridge speciality; it is responsible for Sir Peter Scott visiting the area in 1945 to search for it, and when he found two of them among the European White-fronted Geese. It used to occur annually, but sadly has become very rare in Europe and, as a result, very rare here. The last record was in 2002 at Slimbridge, but it has been recorded at other sites.

Canada Goose (smaller races)

It has a variety of species/subspecies and is one of the most interesting and confusing species in the world. Smaller races occur regularly at WWT Martin Mere and Caerlaverock, and are easily overlooked. The latter site has had four different individuals occurring in recent winters. No need to show pictures, because they all look like small Canada Geese.

Red-breasted Goose

A strikingly marked bird that has been seen at Slimbridge, Caerlaverock and Martin Mere, but again has declined in Europe. It used to occur among any goose species, but is more often seen among Dark-bellied Brent Geese these days which do not winter at any WWT reserves in numbers, thus reducing the chance of a sighting.

Egyptian Goose

This goose was introduced to the UK as an ornamental species and it's found a stronghold in East Anglia. It is spreading from that part of the UK and does turn up from time to time on WWT reserves. It is a brash, loud and obvious bird with a chuffing steam-engine call.

And now, turn the page for the goose-like bird that isn't a goose

Shelduck

Scientific name: *Tadorna tadorna*

Approximate size: 62cm

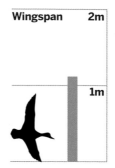

Wingspan 2m

1m

We've looked at the geese, and next up will be the ducks. But first of all, we need to look at an in-betweenie. It's the size of a small goose, but it's not one. It's bigger than any other duck you'll see, too. It's the Shelduck, and even though it seems to sit between two stools, you'll soon find it unmistakable.

Now, you can't tell me that it's not a handsome bird! It's very hard to miss, with its white, chestnut and black plumage, bright red legs and red beak. The male has a knobble of flesh at the base of the bill, and is a bit bigger than the female. They are the largest of all the ducks and the only wild duck with a red beak, so that makes it very easy to pick out.

They are often seen in pairs and, because they are native to Britain, can be seen all year round at wetland centres such as WWT's. They often nest far inland and can have up to 20 ducklings.

Where and when?

MARTIN'S QUICK QUIZ
Shelducks nest in holes or rabbit burrows. True or false? Answer on page 251.

SMALL WADDLING BIRDS

Ducks are fantastic! They come in a huge variety of shapes, sizes and colours, and they are wonderfully entertaining to watch. Of all the wetland birds, I think ducks have the most character.

Broadly speaking, we can divide the country's wild ducks into three categories: dabblers, divers and sawbills. We'll kick off with the **dabbling ducks**. How do you know the duck you are looking at is a 'dabbler'? Well, there's no polite way to put this. If it feeds by sticking its head in the water and its bum in the air, it's 'dabbling'. In short, if your duck looks like it is doing an underwater headstand, this is the category to look in. There are seven main species in this country to look out for.

Mallard — The biggest of the dabblers, and the most familiar (p46)

Teal — Rock star 'face paint' yet small and delicate (p49)

Wigeon — Steep forehead with yellowish forehead (p52)

Gadwall — Marbled plumage, and listen out for its whistle (p58)

Shoveler — The bill is more spoonlike than shovel-shaped (p55)

Pintail — Sharp pointed tail and white head stripe running down to breast (p60)

Garganey — Distinctive head stripe stops at neck (p62)

Mallard

Scientific name: *Anas platyrhynchos*

Approximate size: 58cm

Wingspan   2m

1m

These ducks are found all year round in reserves, urban parks and anywhere there is water. A pair once flew into my garden and hung around a bit on a particularly wet March day. They may be common, but they are nonetheless lovely looking birds.

The male has a bottle-green head, yellow bill, mahogany chest and white ring around the base of the neck. Look out, too, for orange legs, pale grey body, dark wings and a black bum.

And even if you reckon you know the Mallard inside out, do take a look at the tail. Males have these two wonderful little black curled feathers, like 'kiss curls' at the base of the tail. A delightful touch.

Where and when?

The female, however, is not as colourful, being almost uniformly brown. She has orange legs, and the tip of her bill is also orange. On her wings are flashes of brilliant blue. Many of the female dabbling ducks look like this, predominantly brown, with perhaps a discreet splash of colour on the wing.

The female Mallard is a bird who relies more on personality than bling

There is a reason for this and it comes down to childcare. Once her eggs are laid, the female will sit on them until they hatch, the male playing no part whatsoever.

So the female has to rely on her camouflage to keep her and her brood safe, while the males can safely swim around, all bright-coloured and gorgeous. Oh, nature can be cruel at times!

The easiest way to establish the species of a female dabbler is to look at the males. Ducks are often seen in pairs, or with a group of males trying to attract the attention of a female, so if you can identify the male, the female will be the one swimming next to him – or being chased by him!

Teal

Scientific name: *Anas crecca*

Approximate size: 36cm

From the biggest dabbler, which was the Mallard, to the smallest, the Teal. He may be little, but what he lacks in size he makes up for with 1970s eye make-up.

Yes, the male Teal is the Ziggy Stardust of the duck world. He has a dark red head with a broad stripe of dark green, edged by a fine yellow line across his eye. His body looks pale grey – but have a closer look through binoculars. Each feather is edged with a thin black line, giving almost a 'marbled' effect.

He has a green flash on his wing and a white 'go-faster' stripe running across it. The female is the speckly brown you'd expect, but she also has a lovely bright green flash on her wing and a pale-yellowish patch next to it. Both her bill and legs are dark.

Teal are seen all year round in this country, although there are fewer around in summer. The numbers swell from August to April when migrating Teal come in from Scandinavia and Russia.

In the daytime, Teal are often seen hanging around with another species of dabbling duck, the Wigeon, a bird that we're going to meet next.

Wingspan 2m

1m

Where and when?

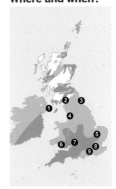

MARTIN'S QUICK QUIZ

Sometimes a Teal has a vertical white stripe dropping down from its shoulder. What does this mean? Answer on page 251.

The male Teal has two very distinct triangles on either side of the undertail

Wigeon

Scientific name: *Anas penelope*

Approximate size: 48cm

Wingspan 2m

1m

As you know, we're in the section that covers 'dabbling' and 'ducks', but actually Wigeon behave rather more like geese, preferring to eat grass out in the fields than feed upside down on water. They are rather dumpy little birds, with short bills, ideal for grazing. They are often seen in large flocks in the middle of a field and if they get spooked by something like a Peregrine Falcon or a Sparrowhawk, they all take to the air together and land back on water. Once they feel the danger is over, they will leave the water and march in a neat squadron back out into the field.

As is fairly standard among the dabblers, the female doesn't have much in the way of identifying features, so let's concentrate on the male. He has a chestnut head, with a mustard-coloured stripe down the front. His chest is a sort of ruddy pink, the body is grey and he has a very distinct white patch on his forewing, and a white belly.

Wigeon are seen across many parts of the country when they come from their Arctic nesting grounds in October, and they will stay on until March.

If you see a tightly packed flock of ducks grazing in the middle of a field, they will be Wigeon

Where and when?

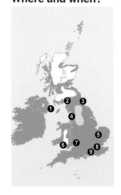

MARTIN'S QUICK QUIZ

Sometimes you will see a Wigeon, like the one shown here, with very different colouring on its head. What is it? Answer on page 251.

Shoveler

Scientific name: *Anas clypeata*

Approximate size: 48cm

Sitting in a hide in the North East of England with Bill Oddie once, I noticed a duck that was swimming in a most peculiar way. 'There's a duck out there that looks like it's dropped its contact lenses in the bottom of the pond,' I said to Bill. He didn't even need to look. 'It's a Shoveler,' he said. And, naturally, he was right.

When a Shoveler hasn't got its face in the water you can tell what it is straight away because of its extraordinary, outsize beak, that looks, unsurprisingly, not unlike a shovel, or perhaps a big spoon.

That bill works the same way as a flamingo's, or indeed a baleen whale's mouth; inside is a series of sieves and bristles which holds onto the food it collects by filter feeding, while letting the water drain out. Shovelers often swim in circles, exactly as if they're looking for something, but what they are doing is stirring up food from the bottom. They will up-end to feed, too, but sieving away at the surface is their preferred method.

The easiest way to identify the female Shoveler (see over the page) is by the shape of the bill because otherwise she does look very like a female Mallard

Wingspan   2m

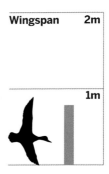

1m

Where and when?

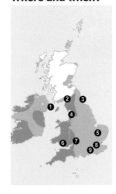

Male Shoveler

Female Shoveler

If the bill isn't visible for you to identify the bird that way, the male is very easy to remember. He has a dark green head like a male Mallard, but with a bright white chest and chestnut body.

Shovelers migrate to the UK from Continental Europe, arriving during October and staying until March.

MARTIN'S QUICK QUIZ

Where do you think Shovelers prefer to feed: shallow or deep water? Answer on page 251.

Gadwall

Scientific name: *Anas strepera*

Approximate size: 51cm

Wingspan   **2m**

1m

Another duck, like the Wigeon, that likes to graze as well as dabble is the Gadwall. Short and compact, these are the least showy of the dabbling ducks; even the male is, well, a bit beige.

See what I mean? But he is another one who benefits from being looked at through binoculars. That pale plumage is more intricate than you might first suspect. He has a black bum, orange legs and a lovely chestnut and white flash on his wing, although often he seems embarrassed to show this bit of embellishment and keeps it hidden.

The female has a flash of white on her wing, too, but her main bit of colour is on her beak, which is bright orange with a dark stripe on the top.

They arrive in Britain from Continental Europe in October and stay until March, although there are breeding populations at several WWT centres if you are reading this in June and can't wait to go Gadwall-spotting until Autumn!

Where and when?

KATE'S TOP TIPS

The male plumage is a mass of fine dark lines like a monochrome Bridget Riley painting

Pintail

Scientific name: *Anas strepera*

Approximate size: 58cm

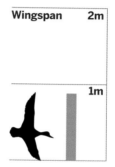

Wingspan   2m

1m

Like the Shoveler, another dabbling duck that has a very obvious, identifying feature is the Pintail. And, like the Shoveler, the clue is in the name. Pintail. Spot it?

The Pintail is known as the 'greyhound' of ducks, because it is sleek and long-necked, and less dumpy-looking than many of the other ducks. The male is undeniably handsome, with his dark chocolate head contrasting with his pale blue bill and white breast. The body is pale, allowing all the drama to happen at the back end; beautifully patterned feathers and the one that gives it its name, tapering to a fine black point. It is this that drives the females wild, and who can blame them?

The female has no pin, and only her sleekness and narrowness tell her apart from her other co-dabblers, although her underside is noticeably pale.

The males will try and find a mate before they leave Britain at the end of March for their breeding grounds in the Arctic and then they will migrate together. This bird is another duck where you will find the female by watching the males. There'll be a lot of flirting and shaking of tail feathers going on.

Where and when?

The white on the breast curves up like a question mark behind the ear

Male

Female

Garganey

Scientific name: *Anas querquedula*

Approximate size: 39cm

Wingspan 2m

1m

The last, but most definitely not the least, of our dabbling ducks is the Garganey. Only just larger than Teal, these little ducks come here in March to breed.

They are quite scarce, only coming to Britain in their hundreds rather than thousands, although they can sometimes be seen in small flocks.

The male's plumage is a hotchpotch of patterns, but he has a very clear pale stripe over his eye, and the rest of his head is dark chocolate brown.

The female's is pale with a dark stripe going through the eye and pale, almost yellow stripes above and below the eye.

She has another dark stripe across the top of her head, a buff throat patch and a pale spot at the base of the beak, known as a loral spot, if you want to show off!

The Garganey, rather sensibly, spends the winter months in Africa

Where and when?

MARTIN'S QUICK QUIZ

One of the trickiest aspects of dabbling duck identification is sorting out the females. As Kate has already mentioned, one of the best things to do is check to see which males they're hanging out with. But in a mixed group, you need a little more to go on.

So, to test your female dabbler skills, I've got seven pictures here, one of each of the females of the species we've just covered. I've cross-referred them to their relevant pages, but see how many you can get right before you check the answers.

Shape is often a good indication – but here's another tip. Each species has what is known as a speculum – it's a flash of colour on the secondary feathers of the wing, which appears on the resting bird as an approximate rectangle near the back of the body. Colours vary, so they're a good indication of which species you're looking at.

p52

p55

p60

p62

p58

p46

p49

SMALL WADDLING BIRDS

Now, from birds who just get their heads wet to birds who disappear under the water altogether.

Watching diving ducks can be a bit alarming. One minute you are following a sweet, perky little duck going about its business, the next it has disappeared entirely without trace. **Diving ducks** do exactly that. They completely submerge themselves in their quest for food. Also, they don't walk; they might sit at the edge of the water, but walking is not something a diving duck does, so if the duck you're looking at is walking, go to another category!

Tufted Duck — Looks black and white from a distance (p68)

Pochard — The rich chestnutty head is most distinctive (p70)

Goldeneye — The bright yellow eye and face patch really stand out (p73)

Tufted Duck

Scientific name: *Aythya fuligula*

Approximate size: 44cm

Wingspan 2m

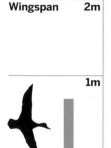

1m

Like the Shoveler, with its distinct bill shape, you can take a pretty informed guess that if the duck you're looking at has a sort of Hare Krishna ponytail hanging off the back off its head, it's a Tufted Duck. Or a Tufty, as they're often known, and as Martin calls them. Even the female has a tuft, although it is a bit more discreet. Both have a yellow eye with dark iris.

As with other diving ducks, the females are less elaborately coloured than the males. Unlike the dabbling ducks, however, at least the females don't all look the same as each other. So that's something of a relief! The male is predominantly black, with a white belly. The female is two-tone brown. Small and rather bustling, these are energetic little ducks that completely submerge when looking for larvae and other underwater invertebrates to feed on. They can be seen all year round on gravel pits, lakes and parks.

Where and when?

The Scaup is a similar bird that can occur among Tufted Duck flocks

The Tufted Duck has a short, scooped beak which is a lovely blue-grey colour

Female

Pochard

Scientific name: *Aythya ferina*

Approximate size: 46cm

Our next member of the diving ducks is the Pochard. The male Pochard is easy to recognise because he is a glorious redhead – auburn rather than ginger – with a blue-grey and black beak and red eyes. His breast and rear end are dark, almost black, and he has a pale grey back and wings.

The female looks like a male who has faded in the wash, so all her colours are more muted and indistinct. Her breast and rear end are more brown than black, her middle section grey, speckled with pale brown, and her head is pale brown, but adorned with a yellowish stripe across the eye.

They spend the winter with us, before heading off to Eastern Europe and Central Asia in March to breed. You will see them up and down the country, where they often feed at night and sleep in big groups, or 'rafts', on the water during the day. They like hanging out with Tufted Ducks, which we've just met, but they look completely different, so luckily there is no chance of getting the two confused.

Wingspan   **2m**

1m

Where and when?

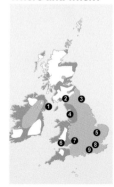

The female's speckles look as if someone has flicked paint at her

KATE'S TOP TIPS

Male

Female

MARTIN'S QUICK QUIZ

The pochard shown below
has a bright red bill. What is
it? Answer on page 251.

Goldeneye

Scientific name: *Bucephala clangula*

Approximate size: 46cm

Our final diving duck is another migrant, arriving here in October from Scandinavia. Rather like the Tufted Duck, its name gives away its chief identifying feature – the gold-coloured eye.

The male has the flashier plumage, with a dark green, almost black head, bright white cheek patch, breast and underparts, and black back. Oh, and very orange legs. Their heads are rather triangular in shape and so is the bill.

The female has a chocolate brown head, a pinkish tip on her beak and the rest of her is dark or pale grey, but she, too, has that startling yellow eye.

These are real speciality divers, spending a long time underwater looking for nice juicy wiggly things. They are often in loose groups on reservoirs and gravel pits during the winter months.

Look out for the Goldeneye's very striking zebra-striped wings (clearly seen on page 67)

Wingspan   2m

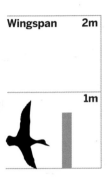

1m

Where and when?

Displaying male Goldeneye

Female

SMALL WADDLING BIRDS

So to our final duck category, and some of the most handsome of the species, the **sawbills**. These ducks mainly prey on fish and their bills have saw-like teeth along the cutting edges.

There are three that you might see in the UK, and as you'll quickly spot, two are rather similar, while the third is quite distinctive.

Goosander — Smooth headed, and found on inland freshwater (p78)

Red-breasted Merganser — More
of a 'punkish' style, and usually
seen near coasts (p81)

Smew — There really
is nothing like it: neat,
compact and a brilliant
white (p83)

Goosander

Scientific name: *Mergus merganser*

Approximate size: 62cm

Wingspan   2m

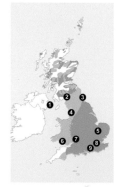

1m

The most commonly seen of the sawbills is the Goosander, which Martin thinks looks like a battleship and I think looks like a Cormorant in drag.

The male has a dark green head and a long, straight, red beak which is hooked at the end like a Cormorant's bill. The rest of its body is predominantly white, but the back is black and the tail grey.

All female sawbills are known as 'redheads', so Martin tells me, and the female Goosander does indeed fit that description. Her predominantly grey body ends abruptly in a sharp line at the neck, and her throat and neck are rich reddish-brown.

Like the male, she has the same long, straight, red bill, hooked at the end. She also has a couple of adornments he doesn't have. Her chin is white, she has a black patch at the base of her beak extending across her eye but, most importantly, a variety of hairstyles, which change depending on her mood.

Where and when?

The Goosander's hunting style makes it look as if it's snorkelling

Male

Female

*Note the Goosander's
hooked bill*

The female Goosander has a sort of long crest, which hangs down at the back of her head, almost as if she is wearing a wig. In her wilder moments she will punk it up a bit, and it will appear more spiky and Mohican-like.

Goosanders are voracious hunters and will eat anything, including big fish. To find their prey they swim with their heads in the water, and as soon as they spot something, they dive for it.

MARTIN'S QUICK QUIZ

In terms of habitat, what's the best way to tell the difference between the Goosander and the Red-breasted Merganser? Answer on page 251.

Red-breasted Merganser

Scientific name: *Mergus serrator*

Approximate size: 55cm

Wingspan   2m

1m

Now, if you were along the coastline and you saw a bird that looks very like a Goosander, take a closer look. It will most likely be a Red-breasted Merganser.

These are slightly slimmer, sleeker and less battleship-like than the Goosander, but let's compare them directly and play 'spot the difference'.

The Merganser's bill is a different shape, narrower and straighter, and without the prominent hook on the end. The eye is red and he has an unruly crest. Overall, his body is darker, with less of the stark contrast of the Goosander's black and white.

The females are probably rather harder to tell apart. Again, as with the males, the shape of the beak is completely different and the female Red-breasted Merganser also has a red eye.

She, too, has the classic female sawbill 'redhead', but the colour merges down the neck and into the pale grey body, so you don't see the very distinct line between head and neck that you do on the Goosander.

Where and when?

The male looks rather like he's been attacked in the playground with a pair of scissors

Male Red-breasted Merganser

Female Red-breasted Merganser

Smew

Scientific name: *Mergellus albellus*

Approximate size: 41cm

Our third sawbill looks completely different, and in fact, barely looks like a sawbill. It is the wonderfully named and equally wonderful-looking Smew.

These are small, compact little ducks, and the pattern of their plumage, particularly that of the males, is so delicate and geometric that they look as if they've been hand-painted.

The males are known as 'white nuns', although I think the black bandit-style mask over their eyes slightly diminishes their nun-like appearance. But the head, neck and breast are startling white and they have a tufty sort of forelock, which can be fluffed up to a crest-like appearance when they are trying to attract a female.

The bill is short, blueish-grey and more triangular in shape. Then there are these wonderfully sweeping lines of black down the back of the head, over the forewing and down the back.

The female Smew has a slight bandit mask, but not as prominent as the male's (see over the page)

Wingspan 2m

1m

Where and when?

In common with her other sawbill counterparts, the female has a red head sitting on an almost uniformly grey body. She has a white patch under her chin, like the Goosander, but the bill, and indeed the bird in general, is such a different shape you would never confuse them.

They breed in Scandinavia and turn up in Britain right towards the end of the year, usually in December, and stay just until early March. They don't come in huge numbers and are rather shy little birds, but worth the effort to try and see. If you're London-based, the Wetland Centre at Barnes is a good place to try.

Female

Not seen it yet? Perhaps you've seen a duck that doesn't quite look like any we've covered so far? There's a good chance it'll be on these pages that Martin has put together.

Mandarin Duck

What an amazing plumage: it should make it so obvious, but the Mandarin prefers to loaf in the shaded cover of overhanging trees on lakes, rivers and streams, and can be difficult to see, as it perches up in the branches. The Mandarin Duck has been introduced to the UK from the Far East.

Ferruginous Duck

Although still a rare duck in Western Europe, they do occur widely, but locally, around the Mediterranean Basin especially. This chestnut-coloured duck has been seen on a number of WWT reserves where Pochard occurs. The male has a pale eye, the female a dark one but both sexes show gleaming white undertail. They migrate with Pochard in the autumn so they can show up among its much larger relative.

Ring-necked Duck

This American diving duck was added to the British list in 1955 and was actually discovered at WWT Slimbridge by Lady Phillipa Scott. It has occurred a number of times since and at other sites, too. The male is very much like a Tufted Duck, but with a peaked head, clear white band on the bill and grey flanks with white section nearer the breast. The female also shows the peaked head and has paler markings about the face and a clear eye-ring and line behind the eye.

Eider

A large stocky and long-billed sea-duck, which can turn up on estuaries, particularly during migration times and during or after gales. A very distinctive duck, it has a call which I always think sounds like the Monty Python women oohing and aahing at each other.

Common Scoter

Males and females are dark birds, although the male is a glossy black all over with a swollen bill that shows a bright yellow patch on the top. It also has a thin yellow eye-ring. The female is a brown bird with paler cheeks and neck, and can be confused with other duck species. Windblown migrants in estuaries or birds lost after mist and fog can turn up anywhere and do so alone or in small flocks.

Long-tailed Duck

This exquisite and elegant sea-duck may turn up occasionally, although not usually in breeding adult form, but in winter plumage or as young birds.

Great Waterbird Challenge
Part 2: Everything counts

We'd already got 15 ticks from the hide, when I spotted what looked like a very tiny duck slipping under the water with barely a ripple. I pointed towards where it went under. 'Another diving duck?' I asked Martin. 'Ah, not everything that swims is wildfowl,' he countered. A few seconds later, he nodded in a direction much further to the left than I was looking. The bird had popped up again several yards away. 'It's a **Little Grebe** (p95),' he whispered.

It really was very small indeed, as it busily scanned the water, before diving once more in a graceful arched movement that would have scored good marks for execution at the Olympics. 'There are quite a few other swimming birds to look out for, in addition to the members of the wildfowl family', Martin continued. 'The easy ones are the **Coots** (p102) and **Moorhens** (p100). At a distance, both look principally black, but the Coot has a white patch on its forehead, whereas the Moorhen's is red. You're also more likely to see a Moorhen foraging around on the grass near a bank – look, there's one now.' There was indeed, and it was scuttling around near the Wigeon, looking like a long-legged sooty hen.

With our total now up to 18, and nothing else in evidence, I wondered if it was time to move on. 'Let's just spend 10 minutes or so counting up the birds you might easily miss,' Martin suggested. I asked what he meant. 'Well, for example, have you seen a Crow yet?' I thought about it, and suddenly realised I wasn't sure – I'd been spending all my time so far gazing

Glorious Great Crested Grebes, in their breeding plumage

Coot

Moorhen

Little Grebe

at the water. 'If we're going to reach our 100 species,' he went on, 'we need to make sure we sweep up all the easy ones, too.'

He was, of course, absolutely right. We raised our eyes to the treeline, where a couple of Crows were lazily wheeling around. A foraging party of tits – Blue Tits, Great Tits, Long-tailed Tits – were peeping and flitting their way through the late foliage, and a Blackbird was chuckling nearby. These are the birds that we're all familiar with from our parks and gardens at home, and as they're not particularly wetland specialists, we're not including them in this book. But they're found in all sorts of habitats, and you'll need to keep count of them in your quest for the big One-Zero-Zero. About 10 minutes of watching gave us an astonishing 15 in all – Robin, Dunnock and Woodpigeon included – taking our tally up to 33.

Just as we were about to move on, number 34 appeared. Around a corner of the lake we were watching swam an elegant long-necked bird. '**Great Crested Grebe** (p92),' said Martin. I frowned. I'd seen Great Crested Grebes before – I've watched them performing their extraordinary breeding display several times – but the birds I was familiar with had glorious plumes, like the manes of lions, adorning their heads. This bird had no such finery. 'Winter plumage,' said Martin. 'Rather like the ducks, which lose their breeding plumage during a period we call the eclipse, the grebes adopt a winter look, too.'

So, a third of the way towards our target, we decided to set off for the estuary, where Martin was confident that several species of waders were awaiting us.

It wasn't even 9am yet, and my hopes were very high. I took a swig from my thermos, gave the water one last scan, and off we went.

Now turn the page to find out more about the Coots, Moorhens and grebes - the other swimming birds.

OTHER SWIMMING BIRDS

We've had a good close look at the wildfowl of the UK, and by now you should have a pretty fair idea of the swans, geese and ducks that you'll see swimming on our lakes and scrapes. Thing is, they won't be alone. There are a few other birds that you could well see on the water's surface, and I'm going to round them up next.

The main species that you'll see will probably fall into one of two families – the grebes or the rails. We're going to cover here the five that you're most likely to see.

Great Crested Grebe – Magnificent mane and courtship dance (p92)

Little Grebe – Delicate body with flattened appearance (p95)

Moorhen – Perky, jerky head movements and red bill (p100)

Coot – Aggressive customer with white bill and faceplate (p102)

Water Rail – Skulker with long bill and pencil-grey face and underparts (p104)

Great Crested Grebe

Scientific name: *Podiceps cristatus*

Approximate size: 48cm

Wingspan   2m

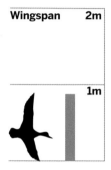

1m

I can't help feeling that this is a bird whose name does it no justice at all. The ornamental feathers that adorn the head of the Great Crested Grebe are far, far more than just a crest. They're a veritable mane, a complete leonine set of chestnut and black plumes that, during courtship, fill out and frame the bird's face with magnificence.

And what a courtship it is, too. If you're lucky enough to see the mating dance of the Great Crested Grebe, then you'll see something that almost defies the laws of physics. The intended pair – and they're both blessed with that tremendous headgear – spend several minutes shaking their heads at each other, diving under the water, swimming towards each other and away, and all the while preparing for the grand finale. Grabbing vegetation from below the water's surface, they present it, then swim off some distance apart. Like two duellers, they then turn and face each other, lower their heads and hurtle together at terrific speed. Just as it looks as if they're about to collide with a sickening thud, they raise their bodies out of the water, pedal furiously on the water's surface so that they're effectively standing on it, and for what seems like an age, face each other bill to bill while violently vibrating their botanical gifts. Once it's over, you're left breathless!

Where and when?

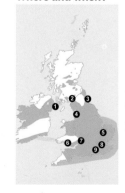

Great Crested Grebe, non-breeding plumage

During the winter, these grebes lose their wonderful plumage, and are left with white faces and cheeks and just the remnants of the crest.

Watch grebes on land, however, and it's a different story. With legs far back on their bodies, enabling that powerful swimming action, they're pretty rubbish walkers, waddling along like drunken penguins looking for a lamp post to lean against. As they spend virtually all their time on the water, only emerging onto small islands of land to build their nests, their ambulatory shortcomings don't hinder them too much.

With regular sightings on gravel pits, lakes and the like, they're easy birds to add to your list, but ones you'll keep coming back to watch time and time again.

Little Grebe

Scientific name: *Tachybaptus ruficollis*

Approximate size: 27cm

The first thing that strikes you about the Little Grebe is the rather unusual shape of its body. It's a small bird, true, but unless it's fluffing out its feathers, it can have a very flattened look, as if someone stepped on it. However, if its feathers are indeed fluffed up, for display purposes, then its rear end looks rather like a delicate powder puff.

Emerging from its circular patty of a body is a short neck, dark copper in tone with a black cap. The eye is ringed with yellow, and a white flick of a stripe adorns the base of its beak.

Little Grebes can be tricky to follow, as they dive underwater to feed, or escape your gaze, emerging many metres away, often in a reedbed or by the bank of a river, where their dark bodies are well camouflaged.

Unlike the other grebes, you won't find it confusing to identify during the winter months, because it's so much smaller than the rest.

Little Grebes have a trilled breeding call that sounds like the whinnying of a tiny horse

Wingspan 2m

1m

Where and when?

MARTIN'S QUICK QUIZ

What's the other name frequently given to the Little Grebe? Answer on page 251.

Adult Little Grebe and chick

Too small to be a Great Crested Grebe, and too large to be a Little Grebe? We've got three other grebe species that visit this country, and in the summer months their plumage is a delight to behold.

Red-necked Grebe

A few spend their summers here, mainly in southern England. A few more turn up during the winter, when their thick necks help differentiate them from Great Crested Grebes.

Non-breeding plumage

Black-necked Grebe

Again, the south of the country is the best place to see these gorgeous grebes with their golden plumes. A sloping forehead is your best identification point in winter.

Breeding plumage

Slavonian Grebe

Possibly even more sumptuous than the Black-necked, this bird, also known as the Horned Grebe, has a rufous neck and fuller plumes. White cheeks and white foreneck help identify the winter bird.

Non-breeding plumage

Moorhen

Scientific name: *Gallinula chloropus*

Approximate size: 34cm

You're actually just as likely to see a Moorhen wandering near the banks of a water body as swimming on top of it. If you do, look out for a remarkable set of toes – their great length enables the Moorhen to cross floating vegetation, spreading out their weight over a wider area.

The red beak with yellow tip, and red facial shield are the best way of distinguishing a Moorhen from its close relative, the Coot, although streaks of white along the wings and a white fluff at the end of the tail also stand out.

The name, however, is only really half right. When you see Moorhens on land, they really do appear a bit like chickens as they scratch and scrabble out, their heads pecking away while their plump bodies toil along behind. The 'moor' part of the name is misleading, though: you won't find these birds on moorland, but in wetlands of all types. The 'moor' prefix is probably actually a contraction of 'marsh'.

Young Moorhens are browner than their parents, and don't yet sport the distinctive red face shield

Wingspan 2m

1m

Where and when?

Coot

Scientific name: *Fulica atra*

Approximate size: 37cm

You've already discovered that the Moorhen has a red beak and facial shield, and you can now see that the Coot has the white equivalent. What's the best way to remember which is which? Well, we've got an old phrase we can fall back on here – 'bald as a coot'. (Interestingly, the word actually comes from an ancient English word, 'balled', meaning white, rather than featherless or even hairless.)

The faces apart, you can also distinguish the Coot because the rest of its plumage is entirely black, without the splashes and tufts of white displayed by the Moorhen. Rich red eyes are stand-out features, too. Again, unlike the more land-based Moorhen, the Coot has partial webbing on its toes, to aid swimming.

Coots are aggressive birds, setting up all sorts of crackles, trumpets and explosive calls day and night, particularly during the breeding season. You won't tend to see them fly very much, but if you do, watch for their comical take-off. They give themselves a good runway of open water, and trundle along with much splashing and gritted determination before finally taking off for what is almost always an extremely short distance. Lots of energy for very little reward.

Wingspan   2m

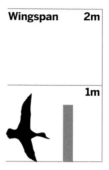

1m

Where and when?

Coots bob their heads when they swim, and frequently dive under the water, too

Juvenile

Water Rail

Scientific name: *Rallus aquaticus*

Approximate size: 26cm

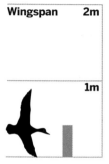

Wingspan   2m

1m

Rather like the Bittern, the Water Rail is a real skulker, and your best chance of seeing it is usually during cold weather when it emerges onto the ice. Look out for the long down-curved bill with red base, lead grey face and breast, black and white striped flanks, chestnut upperparts and creamy white undertail, which is held up and flicked as it walks. You're more likely to see it near the water's edge than swimming, but as it's related to Moorhens and Coots, it sits comfortably in this section.

The Water Rail is one of the actors of the waterbird world, adopting a different role depending on the situation. Sometimes it will freeze like a Grey Heron, standing motionless and hoping noone can see it, ironically giving you the chance of a very good look. Other times it will walk slowly, with rear end held high, like a Moorhen. Then there's its Bittern stance (page 177), tall with elongated neck. Then again, it might act like a Coot, dashing across a ditch.

Finally, Water Rails have one further impersonation: garden birds. If you're at a reserve that puts feeding stations out near water bodies, keep an eye out for the occasional Water Rail that might emerge to tuck in to the bounty provided, alongside the various finches and buntings.

Where and when?

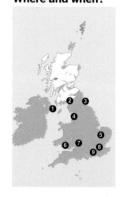

If you hear a pig squealing in the reeds, it's not a pig. It's the Water Rail's call

Great Waterbird Challenge
Part 3: It's all in the detail

W e'd made our way out towards the Severn Estuary on the outskirts of the Slimbridge reserve. 'Right then,' said Martin, 'it's time for the waders. As it's autumn, this is an excellent time of the year to see them. Several species breed in northern climates, migrating south for the winter, and the habitat here is a perfect stopping-off point for them on the way. We might see one or two unusual species mixed in with the regulars. And as it's still nice and early, they should all be congregating around the scrapes where they'll have roosted overnight.'

He was absolutely right. Once we'd settled ourselves at a good viewing point and cast our binoculars over the assembled flocks, our first dozen wader species had already been ticked off. There was a large flock of **Dunlins** (p149) busily jabbing away like sewing machines at the edge of the mud, while several **Curlews** (p113) wandered sedately around an adjoining pasture, looking for all the world as if they owned the place, which, I suppose in some senses they do. **Redshanks** (p129) were everywhere, standing up to their knee-joints in the water, and tiny **Ringed Plovers** (p165) mingled with **Lapwings** (p134) on one of the raised islands in the middle.

'We're in luck,' whispered Martin, nodding towards the Dunlin flock. 'About eight or nine birds in from the left, see that slightly smaller one? It's a **Little Stint** (p152).'

This is why it's so good to go birdwatching with a master of the art. I'd been watching the Dunlins, and hadn't even noticed that one was different. Now that Martin had pointed it out, though, I could see that its bill was slightly shorter and straighter than those of the Dunlins, and the patterning on its back was, well, a bit fancier. 'Always worth scanning a flock of anything,' he continued. 'Some species are sociable, and gather in large numbers like the Dunlins. Others, like Little Stints, although they can be seen in small flocks, often travel around as individuals. From the stint's point of view, burying itself in a flock of Dunlins is a good way of protecting itself.'

I was learning already and, before long, I'd made a discovery of my own. 'See those Curlews on the right of the pasture?' I whispered. 'There's a smaller one in the middle. A juvenile?'

Martin focused in. 'Ooh, good spot,' he said. 'No, it's not a juvenile, it's a **Whimbrel** (p115). Whimbrels are like smaller versions of Curlews, and that Whimbrel you found

was standing next to a Curlew, so you could see the contrast, but it's not always that easy. The best thing to do is look around the bird's immediate landscape – perhaps there's a plant or fence post nearby that can give you an indication of size.'

Time was beginning to tick by, but I was in no hurry to leave the peaceful landscape that stretched in front of me. The 'bird' part of 'birdwatching' was coming along nicely, but a day like this isn't just about lists. The 'watching' part is what it's really all about, and so we spent another hour or so just taking in the behaviour of these wonderful creatures. I came away with a favourite, too. A **Common Sandpiper** (p159) came whipping around the back of the scrape, and I watched it mesmerically. Active yet furtive, it would cover the ground with a few steps and a short flight, and each time it landed, it would bob its body half a dozen times, first head, then tail, the bobbing eventually slowing to a halt. It was like watching a seesaw with wings. Step, fly, bob. Step, fly, bob.

Martin nudged me. 'Brilliant,' he murmured. Red-necked Phalaropes visit Slimbridge only occasionally and, in fact, there had only ever been four records there. We were looking at the fifth.

Martin was delighted. The bird had been around for a few days, and he'd been hoping it would still be present. We both watched the phalarope, as it delicately swam around (very unusual for a wader) looking for food. It's not included in this book, as it's such a rarity, but I mention it here as an example of the unexpected possibilities that a day's birdwatching can tantalisingly offer. The phalarope was one of no fewer than 18 wader species we were able to see and, all in all, our tally had reached 56.

Over halfway there, it was the perfect time to stop for a spot of lunch.

Now turn the page to find out more about the wading birds.

WADING BIRDS

We now come to what is one of the most fascinating categories of them all – the waders. This is a wonderful set of birds, all adapted to feed in different conditions, or to find their food in different ways within the same habitat.

So, for example, it is possible to see a Little Stint – which is smaller than a sparrow – feeding alongside a Curlew whose body is roughly the size of a Mallard. The massive long down-curved bill and long legs of the Curlew contrast hugely with the stint but they can feed on the same estuary mud, because each has a different bill that reaches for different types of prey.

The long, downcurved bill of the Curlew

Most of the birds that you'll find in the pages ahead are migrants passing through – in fact of the 28 species covered only seven actually breed in Britain. Many spend the winter in large flocks and allow other species to 'hang out' with them.

The delicate, upcurved bill of the Avocet

Yes, you read that right: 28 species! That sounds like one heck of a lot of birds to try to sort out from each other as they peck and strut and probe their way across the mud or through shallow water. Unlike the wildfowl we covered earlier into the book, they don't divide into neat groups like swans, geese and ducks, so to help you narrow down which birds you're looking at, we're dividing our wading birds by size.

The bright, probing bill of the Redshank

● If your bird is at least 35cm in length – that is, it has the approximate body size of a duck – then turn to page 110.

● If your bird is between 25cm and 35cm – that is, it has the approximate body size of a dove – then turn to page 124.

● If your bird is smaller than 25cm – that is, it has the approximate body size of a blackbird or smaller – then turn to page 144.

The strong, slightly curved bill of the Dunlin

Large Wading Birds

We're starting our section with the big-bodied waders which, paradoxically, form the smallest group. These are the birds whose bodies are approximately the size of a duck, and they've all got impressively long and, in some cases, curved bills, too. The minimum size for this group is 35cm in length, and there are just six species in all that you're likely to see.

Curlew — Huge, with long, down-curved bill (p113)

Whimbrel — Similar, but smaller than the Curlew; darker markings (p115)

Oystercatcher — Black and white with thick orange bill (p116)

Bar-tailed Godwit – Slight
upturn at bill end (p123)

Black-tailed Godwit –
Long, straight bill and
orange-red plumage (p121)

Avocet – Black and
white with delicate,
upturned bill (p118)

Curlew

Whimbrel (p115)

Curlew

Scientific name: *Numenius arquata*

Approximate size: 55cm

Wingspan   **2m**

1m

The Curlew and the Whimbrel are two very similar species, and the largest waders you will see in Britain. We'll start with the bigger of the two – and the one you've probably already heard doing voiceovers on TV. The Curlew is named not for its curved bill, but in imitation of its call, and the distinctive *curl-ee curl-ee curl-ee* is much beloved of TV sound editors whenever the main characters are out on the moors or near estuaries. It's the call of the wild, and you'll never forget it.

The sheer size of the Curlew helps it stand out, but the overall effect is of blackish-brown feathers, with paler head and neck, and dark streaks running down the breast. See them through binoculars, and you'll notice that those streaks turn into chevrons on the sides, rather like the keep-your-distance indications on some motorways. The bills really are very long, and curve more than the Whimbrel's (easy to remember that – the more curved, the more a Curlew). Interestingly, it's actually the females that have the longer bills of the two sexes.

Broadly speaking, if you see a large wader with a downcurved bill during the winter months, it'll probably be a Curlew, as Whimbrels are very rare this time of year.

Where and when?

Curlew

Whimbrel

MARTIN'S QUICK QUIZ

Why does the Curlew have such a specialised bill? Answer on page 251.

Whimbrel

Scientific name: *Numenius phaeopus*

Approximate size: 41cm

Whimbrels can be seen during the summer months, although they don't often breed in this country, and when they do, it tends only to be in the very north of Scotland. Your best bet is to see them on migration in spring and autumn, as they make their way to and from Scandinavia and Africa.

Martin tells me he has seen flocks of up to 100 resting in estuaries or sometimes fields, and if you're lucky, your Whimbrels will be in the company of Curlews, which gives you a good chance to compare the two.

Overall, the Whimbrel is a darker bird than the Curlew, and it looks as if it's wearing a dark cap when viewed from the side, thanks to the hint of a pale stripe above the eye.

It is shorter billed and legged than the Curlew and in flight it is also noticeably darker. Whimbrels have darker lines running through their eyes, too.

Whimbrels make a distinctive fast whistling call as they fly past

Wingspan 2m

1m

Where and when?

Oystercatcher

Scientific name: *Haematopus ostralegus*

Approximate size: 42cm

The 'Oyc' is one of a few unmistakable birds within the wader group, and that's saying something! Roughly the size of a Woodpigeon, it sports a pied plumage, black above and white below, making it look as if it's wearing a dinner jacket over a white shirt. The head, back, wings and tip of tail are all black, and the breast, belly and undertail are white. Younger birds have a browner tinge to the dark plumage and in their first year they develop a white cut-throat marking.

It is a stocky bird with a striking, long, thick, orange bill and thick, pink legs. In flight they show a stark white backside, which meets in a point on the back, and a broad white band across the wings. Oystercatchers are very active, vocal birds and highly entertaining to watch, especially when breeding. Displays, territorial disputes, protection of the nest from predators, feeding their young: all are done with as much fuss and noise as possible. If you're looking for the rowdiest of the waders, you've found it.

The young are very well camouflaged and hide among stones, waiting for their parents to return. In fact, they are the only wader species we cover in this book that carries food to the chicks even after they have fledged – all other wader chicks fend for themselves and search out food items.

Wingspan **2m**

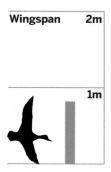

1m

Where and when?

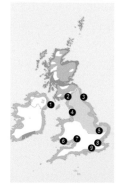

KATE'S TOP TIPS

Oystercatchers have bright, berry-red eyes

Avocet

Scientific name: *Recurvirostra avosetta*

Approximate size: 44cm

Like the Oystercatcher on the previous pages, the Avocet is another pied and unmistakable wader. At first glance you might think that it's a far more delicate and graceful bird – get to know the Avocet, however, and you'll soon change your mind.

This is one of the few wader species that breeds in Britain, and during the months of spring you can have a lot of fun watching its more aggressive side. Pairs will attack intruders relentlessly, and can completely dominate scrapes even when just a few pairs are nesting. They're not as rowdy as the Oystercatcher, but you wouldn't want to mess with them.

Of course, the main feature that stands out is the very fine upturned bill which it uses to sweep side to side as it strides through the shallows. The plumage is largely white, but the Avocet has a black cap and hind neck, with a large black oval on the wings and back when at rest. During this time, it often tucks its bill into its wings. The legs are very long and bluish in colour.

Avocets are one of the few waders that sometimes swim in deeper waters to feed

Wingspan   2m

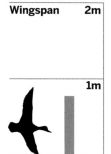

1m

Where and when?

MARTIN'S QUICK QUIZ

Where would you be likely to find an Avocet's nest? Answer on page 251.

Bar-tailed Godwits,
summer plumage (p123)

Black-tailed Godwit,
summer plumage

Black-tailed Godwit

Scientific name: *Limosa limosa*

Approximate size: 42cm

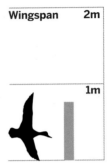

Wingspan 2m

1m

Any large straight-billed wader on an estuary or wetland will almost definitely be a godwit. The only question is: which one? We have two species in this country, both of which are shown opposite in their breeding plumage.

We'll start with the Black-tailed Godwit, or 'Blackwit', as Martin calls them. (See, I'm picking up the lingo already.) You're most likely to find Blackwits feeding on scrapes and wet fields in spring and autumn, but they do spend the winter here – or at least attempt to as long as the ground doesn't freeze too much. Blackwits feed singly or in large flocks using a 'sewing machine' action as they repeatedly pierce the mud, often with their heads underwater and belly-deep.

So how do you tell them apart from the Bar-tailed Godwits, or 'Barwits'? For a start, the bill is noticeably upturned on the Barwit, and somewhat shorter, too. The Blackwit doesn't just have a longer bill, it has longer legs, too.

The plumage is another good indicator – the Blackwit has an orangey-red colour on its breast and neck in summer, plus bars on its belly, whereas the Barwit's rich red colour scheme runs right down the belly, too. But both birds are grey and white in winter. If you're still unsure

Where and when?

which godwit you're looking at, then it's time to play the patience game, and wait for it to take flight. This is the clincher! Blackwits have a very obvious white rump, with a solid black tail and lower back. They also have very obvious black wings with a clear white band. On the Bar-tailed Godwit, as the name suggests, the tail is – well, barred. It also shows a white wedge that runs from the rump, down to a point on its back, while the wingbar is not as clear as on the Blackwit.

In flight, the feet of the Black-tailed Godwit project much further than than those of the Bar-tailed Godwit

Bar-tailed Godwit (winter)

Black-tailed Godwit (summer)

Bar-tailed Godwit

Scientific name: *Limosa lapponica*

Approximate size: 38cm

The Bar-tailed Godwit is a high Arctic breeder and a very long-distance migrant. Thousands occur around the British coast during the winter months, favouring estuaries and sandy shores. Keep an eye out for them during spring and autumn passage, too.

For breeding plumage comparisons with the Black-tailed Godwit, see the previous pages. Out of the breeding season, look out for the more streaky feathers of the Barwit compared to the marbled effect worn by the Blackwit.

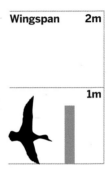

Wingspan 2m

1m

Where and when?

MARTIN'S QUICK QUIZ

The Barwit sports a longer stripe above its eye in winter than the Blackwit. What is the name of this stripe? Answer on page 251.

MEDIUM WADING BIRDS

The middle group of waders, in terms of size, come in at around the scale of a dove, (although generally with longer bills and legs, of course). These are the birds between 25cm and 35cm in length, and they're quite a mixture. We've got some of the plovers here, as well as the very different members of the shank and snipe families, so there's plenty of fascinating variety to enjoy.

Golden Plover — Rich golden hue, often found in large flocks (p137)

Ruff — Unmistakeable when breeding, anonymous during winter (p126)

Greenshank — Pale with light coloured legs and slightly upcurved bill (p132)

Snipe — Very long bill, streaked plumage (p140)

Redshanks — Two lanky species with red legs (p129)

Grey Plover — Metallic grey plumage, usually seen on the coastline (p139)

Lapwing — Familiar bird from fields with prominent crest (p134)

Ruff

Scientific name: *Philomachus pugnax*

Approximate size: 25cm

Look at that second part of the Ruff's scientific name. If you ever get the chance to see male Ruff in full breeding display, you'll see just how pugnacious they are. Male Ruff don't just ooze testosterone towards each other, they even put a ring around their display ground and sell tickets.

The arena is called a lek, and unlike many species in which male displays to female, with the Ruff, males face off against each other, with lunging, jumping, crouching and all sorts of other poses designed to intimidate, It's all backed up by plumage, and the male Ruff, or at least those at the alpha male end of the spectrum, are adorned with tufts, frills, ruffs and all manner of outrageous feather configurations, all symbolising strength and beauty.

Outside the breeding season it's a different matter. The plumage can be patchily variable, and the legs can be greenish, yellowish or even reddish. Males are taller than females, too, so a group of Ruff together in the winter can end up looking like a number of species.

If you have good binoculars, check the Ruff's bill: it 'droops' very slightly at the end

Wingspan   2m

1m

Where and when?

Juvenile

Male, Summer, without ruff

Adult female

In fact, it's a good rule of thumb to say that if you're not sure what it is, it could be a Ruff.

Martin did point out a few handy tips, however. Ruff are always small-headed, and short-billed for their size, and it's worth watching them for a while, because if a slight breeze whips up, their feathers are more likely than other medium-sized waders' to stick up at the back. Females, meanwhile, have black blotches on their scaly looking plumage during the breeding season.

Redshank

Spotted Redshank (p131)

Greenshank (p132)

Redshank

Scientific name: *Tringa totanus*

Approximate size: 28cm

Wingspan 2m

1m

The Redshank is one of those marker birds that birdwatchers like Martin use to make their comparisons. It's present in most places where you find waders, and its size and shape can help you identify other birds even at some distance: a bit smaller than a Redshank; fuller-bodied than a Redshank; longer-legged than a Redshank... you get the idea.

So, the key thing is to get to know your Redshanks. Yes, I use the plural because there are actually two of them, although it's the Common Redshank that is omnipresent.

As the name suggests, the legs are orange-red and the plumage is more clearly marked with darker streaks, chevrons and spots during the breeding season. Ever vigilant, they announce any danger with loud ringing calls. The rest of the time their call is a sort of 'teu' sound... one you'll soon become familiar with as you get to know wetlands, and I think it has a rather melancholy tone.

During the breeding season Redshanks perch on vantage points, such as fence posts or tussocks to look over the marshes for predators

Where and when?

Outside the breeding period Redshanks tend to flock more, particularly when gathering to roost. The plumage becomes much duller and nondescript during this time from July to March.

The Redshank feeds by picking and lightly probing the muddy edges of wetlands or on mud shelves in creeks or estuaries, and here's the first distinction between them and their close relative, the Spotted Redshank. Whereas Spotted Redshanks tend to feed in deeper water, Redshanks are more likely to probe around the shallow edges.

Redshank

Spotted Redshank
(non-breeding plumage)

The Spotshank's bill is slightly longer than that of the Redshank

Spotted Redshank

Scientific name: *Tringa erythropus*

Approximate size: 30cm

Wingspan 2m

1m

This is one of those birds that's almost always abbreviated by birders. The jargon you'll need to know here is 'Spotshank'.

The Spotshank is a very elegant wader, very similar to Redshank in non-breeding plumage, but during the summer months it's quite unique, becoming almost completely sooty black with white notches at the edges of the back and wing feathers and whitish bars on the belly and tail). This stunning plumage is only seen on passage birds on their way to or from their breeding grounds, but it's quite something to see. Most sightings, however, involve non-breeding plumage birds by themselves or in small parties mainly in the autumn, a few in spring and occasionally over winter.

This is the toughest time to tell them apart. In flight, or flapping in the water, the Redshank shows very clear white markings on each wing, which the Spotshank doesn't have. The Redshank also has a clear white wedge on its back which sort of points towards the head, whereas the Spotshank has more of a lozenge shape.

Outside of the breeding season, I think the Spotshank looks rather like a ghostly version of its commoner cousin, with grey upper parts and clean white underparts.

Where and when?

Greenshank

Scientific name: *Tringa nebularia*

Approximate size: 32cm

Wingspan   2m

1m

We've looked at the two redshanks, so guess what the main difference is with the Greenshank. Easy really, isn't it. The legs are indeed greenish, although at distance they look a little more like grey. This bird is larger than the Redshank with clean white underparts, darkish wings and markings on the back of its neck that look rather as if they've been smudged. Overall, I think it has rather a frosted look.

The legs – or shanks – are thicker than those of the Redshank, giving the bird a stockier look. It also looks more thickset in flight, and the wings look very dark against the broad white wedge on its back. From underneath it looks pale grey, but its loud *tew-tew-tew* flight call helps greatly with ID.

The Greenshank breeds in upland areas and islands of Northern Britain but many of the birds seen are likely to be Scandinavian or Russian on passage, sometimes in flocks, but usually in small groups or by themselves. They can also overwinter in this country.

Where and when?

 KATE'S TOP TIPS

The Greenshank feeds by sweeping its bill from side to side, a process Martin calls 'trawling'.

MARTIN'S QUICK QUIZ
Other than colour, what's the main difference between the bill of the Greenshank and those of the two redshanks? Answer on page 251.

Lapwing

Scientific name: *Vanellus vanellus*

Approximate size: 30cm

Wingspan   2m

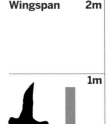

1m

Breeding and wintering in most parts of the country, this attractive wader can be seen throughout the year. In spring, pairs spread out over wetland areas and fields to scrape out their nests on the ground, and for the males to display over them. Flights are made back and forth over the territory swooping and tumbling as they make delightful calls, and throbbing noises with the wings.

When nesting, Lapwing can be seen picking up material and then throwing it over their shoulder into the nest, looking rather as if they're dealing with the superstitious aftermath of spilling salt.

Lapwings are unmistakeable members of this group, as they are the only waders that are crested in all plumages, the crest varying in length as adult males have long wispy crests and juveniles shorter stubbier ones. From a distance the bird can look black above and white below with white markings around the face, but take a closer

Where and when?

MARTIN'S QUICK QUIZ
From its call, the Lapwing was once given another name. What was it? Answer on page 251.

In flight Lapwings often dip and twist as if buffeted by the wind. Look out too for very blunt, rounded wingtips

look and you will see a green tinge and purple iridescence to the plumage. The undertail is an orange-yellow which also stands out. Colouring is similar in winter, but looks more worn.

From June onwards Lapwings form flocks on wetlands to roost, wash and feed. From October to early March they flock in their hundreds if not thousands.

Golden Plover,
breeding plumage

Grey Plover,
breeding plumage (p139)

Golden Plover

Scientific name: *Pluvialis apricaria*

Approximate size: 28cm

We've already had a look at the Lapwing, which is also sometimes known as the Green Plover. Now it's time to paint our palette two other colours beginning with 'G'. The Golden and the Grey Plovers are Britain's main other two medium-sized plovers, and although they're very easy to tell apart during the breeding season, they can be a little trickier in non-breeding plumage.

First up is the Golden Plover, which only really breeds in upland areas during the summer months. They're a bit smaller than the Lapwing, and dressed in a truly beautiful plumage, with solid black from the face to the belly and, as the name suggests, a golden hue to the plumage with fine spangled markings and a large black eye. I think there's something rich and 'treacly' about their colouring.

Come the winter months, we get a much better look at the Golden Plover, as it moves into farmland or wetland areas, often

Wingspan   2m

1m

Where and when?

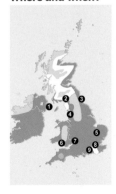

MARTIN'S QUICK QUIZ

When flushed, why do Golden Plovers fly higher than Lapwings? Answer on page 251.

congregating in huge flocks. When feeding they will run a short distance and stop to peck, always looking alert. During the day they will head for wetlands, often in the company of Lapwings, to roost on areas that provide a good view to help detect and avoid predators, especially the Peregrine Falcon that is very fond of eating them. When flushed by a predator they fly faster (more like pigeons) than Lapwings, banding together in a tight flock with synchronised movements, rather like a large flock of starlings.

Golden Plover, winter plumage

Grey Plover, winter plumage

When at rest, a large flock of Golden Plovers almost always all face the same direction, into the wind

Grey Plover

Scientific name: *Pluvialis squatarola*

Approximate size: 28cm

The Grey Plover is a passage migrant and wintering species in the UK, preferring open shorelines such as estuaries and beaches, although it can occur inland on passage and during gales. Although it does occur in flocks, the birds tend to space out when feeding, and are only really found standing shoulder to shoulder at high tide or on small island roosts.

Where the Golden Plover has the rich, warm treacly colouration I mentioned, the Grey Plover has a more steely look about it. If you're lucky enough to catch sight of one in breeding plumage at the edges of its migratory seasons, look out for a pale grey hue to the cap, hindneck, back and wings which, when studied closely, actually reveals a pattern of fine intricate black markings.

Winter is the testing time for plover watching, so here are the key features to look out. The Grey is the stockier bird with a longer, thicker bill. In winter plumage, the Golden retains a hint of its yellowish warm appearance, while the Grey's plumage has a colder tone, and looks as if it's feeling the cold. The Grey also has a more prominent stripe over the eye, and its paleness makes its dark cap stand out more than the Golden Plover's. The Grey also has a more obvious white wing bar in flight.

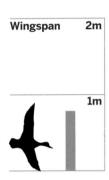

Wingspan   **2m**

1m

Where and when?

Snipe

Scientific name: *Gallinago gallinago*

Approximate size: 26cm

Wingspan 2m

1m

A small sewing machine – that's a pretty good description of the pecking, feeding action of a Snipe. If it's not drilling the ground, you're likely to see it sunning itself on the edge of a scrape or marsh, or perched on a post near its breeding grounds. One thing is certain – if you see one or two or even half a dozen or so Snipe, you can be fairly certain there are many more hidden in cover nearby.

Heavily streaked plumage, and a lengthy bill that seems out of proportion to the rest of the bird, the Snipe is best seen during the breeding season, particularly during evenings when its display flight comes into its own. I reckon that several ghost stories told from misty marshland actually derive from the sound made by feathers in the bird's outer tail that produce a remarkable throbbing sound, known as drumming. Even when you know it's Snipe you're listening to, it's an eerie sensation.

Out of the breeding season, they're much harder to discover. They freeze at the sign of danger, and you may stumble across one unwittingly. Rather as Woodpigeons burst from trees, Snipe burst from the ground with a swishing sound as you walk past, having left their frantic flight to the last moment. In the second or two it takes you to recover yourself, the bird has become a speck in the sky.

Where and when?

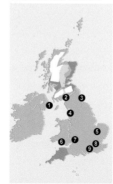

*Snipe frequently
bob up and down
as they feed*

Snipe

MARTIN'S QUICK QUIZ

Another, smaller type of snipe (below) visits Britain in winter. It's shorter-billed and has broad buff streaks along its back and a greenish tinge. It has a very bouncy feeding motion, rather like a jack-in-the-box. Do you know what it's called (there's a clue in the previous sentence)? Answer on page 251.

Woodcock

Scientific name: *Scolopax rusticola*

Approximate size: 34cm

The Woodcock is mainly a bird of woodland, forest and thicket, but as they can appear in the open, in ditches and fields near water bodies when it's snowy, we're giving it a page here to show how it contrasts with the Snipe. Its colouring closely matches the undergrowth of a woodland floor, but it stands out in the snow.

A much larger bird than the Snipe, the Woodcock's breeding population is augmented by visitors from Scandinavia and Russia during the winter months. Like the Snipe, it sits very still as you unwittingly approach, and when flushed flies in a zigzag motion low to the ground. If you do get a chance to see one, notice how far back the eyes are on the head.

Wingspan   2m

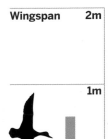

1m

Where and when?

SMALL WADING BIRDS

We've reached the smaller of the waders, which range from sparrow- up to blackbird-sized. These birds are all less than 25cm in length, and several can often be found in each other's company, which all helps for identification. They also include a couple of plovers, a family which you've already met.

Sanderling — Ghostly white, seen along shorelines (p156)

Sandpipers — Bobbing tails and energetic feeding habits (p159)

Ringed Plovers — Stocky with bold black markings on faces (p165)

Turnstone — Bullish little bird with near orange colouring (p168)

Dunlin — Slightly
downcurved bill, and black
belly in summer (p149)

Knot — Huge flocks
in winter, brick red in
summer (p146)

Curlew Sandpiper —
More downcurved bill,
and likely alone or in
small groups (p151)

Little Stint — Short
straight bill, and
absolutely tiny (p152)

Knot

Scientific name: *Calidris canutus*

Approximate size: 24cm

Wingspan   2m

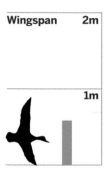

1m

One of the most exciting moments I've ever had birdwatching was when Chris Packham took me to see a mighty flock of Knot. As they stormed over our heads in their masses, their wings whirring so loud you could hardly hear yourself speak, I had an amazing sense of the wondrous maelstroms that nature can provide – an assault on your senses, and utterly breathtaking. All in all, a remarkable performance for a dumpy wader with short legs, that barely reaches the shoulder of a Redshank.

Knot are usually found in huge flocks on large tidal mudflats and sandy beaches during the winter months. They can occur pretty much anywhere on passage around the coast, but more often in the autumn. Although uncommon inland, you might come across a few on scrapes at this time of year.

Where and when?

Known as Red Knot, their breeding plumage is spectacular, making them look as if they've rolled in brick dust. In winter, like most waders, they're much paler, grey above and white below. Look out for pale backsides and underwings as a huge flock flies overhead.

In regular high-tide roosts they become a surging mass, all running together as the tide floods, and jostling for position in the best spots.

Knot have a rather lethargic feeding style compared to other waders

MARTIN'S QUICK QUIZ

If you see a hint of peach on a Knot's breast, what does that indicate? Answer on page 251.

Dunlin

Juvenile Dunlin

Little Stint (p152)

Curlew Sandpiper (p151)

Dunlin

Scientific name: *Calidris alpina*

Approximate size: 18cm

We're into the tiniest little waders now, and your benchmark species for them is the nearly ubiquitous Dunlin. Great little bird, great little name – it's easy to become very fond of the Dunlin.

The first thing to talk about is the bill. As small waders go, it possesses a reasonably long one, which curves downwards very slightly towards the end, as if it was supporting a small weight from the tip, suspended by a piece of string. Remember, the curve really is very slight and gradual, which is important to remember when it comes to comparing it with the Curlew Sandpiper on page 151.

In the breeding season, the Dunlin is unmistakeable. Normally settling in sizeable flocks, it sports a dark patch on its belly, which makes it look as if it stood too close to a candle and got a bit singed. The plumage is a nicely mottled black and brown.

Our commonest small wader, it can be found in the UK all year round, but in the winter it feeds in huge flocks, sometimes numbering in their thousands. The plumage at this time of year has a wintry feel to it, too, with a light grey back fading into a white underbelly.

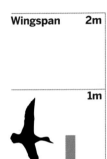

Wingspan   2m

1m

Where and when?

Winter flocks of Dunlins can number in their thousands

Curlew Sandpiper

Scientific name: *Calidris ferruginea*

Approximate size: 18cm

Wingspan 2m

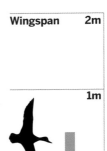

1m

If you're looking at a small wader with a downcurved bill, then it's almost certainly a Dunlin, which we've just covered, or a Curlew Sandpiper. So, how do you know which is which?

For a start, the Curlew Sandpiper's bill is slightly more obvious than the Dunlin's, with the overall curve more evenly spread along the bill. Slightly bigger than the Dunlin, it also has longer legs and tends to feed in deeper water, often up to its belly. This is a good time to look at body posture, as the Curlew Sandpiper feeds by peering down into the water, so its rear end is slightly raised, whereas the Dunlin holds itself more horizontally.

The Curlew Sandpiper is best looked for in the autumn, while on passage from Siberia to Africa (quite a flight for such a small bird).

Where and when?

If you're lucky enough to see one on spring passage, look out for its brick-red plumage at this time of year

Little Stint

Scientific name: *Calidris minuta*

Approximate size: 13cm

Wingspan   2m

1m

We're in the section marked 'stints', so it's about time we actually came across one. Little Stints are truly tiny waders that breed on the tundra of northern Scandinavia and Russia, but not in the UK.

So the only time you're likely to see one is while they're on passage in spring, and more usually in autumn. If they've had a good breeding season, then we might get quite a few visiting during the autumn months.

The key features to look out for here are the very small size, short bill compared to the previous two birds, dark legs and clean white underparts to the breast.

The best place to look for them is on estuaries and at the muddy edges of freshwater wetlands, normally among flocks of Dunlin.

Where and when?

When among other species of small wader, Little Stints stand out by their minute size but the whiter underparts remind me of a Sanderling (p156)

MARTIN'S QUICK QUIZ

The Temminck's Stint is also sometimes found in Britain, and its legs help mark it out from the Little Stint. In what way? Answer on page 251.

MARTIN'S QUICK QUIZ

Here's a photo of three species of wader in one area. Using your growing knowledge of waders, can you identify them all? And do you know what type of gull is with them? Answer on page 251.

Sanderling

Scientific name: *Calidris alba*

Approximate size: 20cm

Wingspan   2m

1m

There's another member of this family that you've got a good chance of seeing in this country, but you'll be glad to know it's quite distinct. The Sanderling is the 'white ghost' of the waders, and prefers sandy beaches and mudflats, where it runs up and down the tideline staying just in front of the lapping waves.

I say 'ghost' because its white plumage, as it scampers up and down a chilly coastline, puts me in mind of some tiny paranormal being. Step back a bit further, however, and their collective bodies look a little like foam at the edge of the tide. As you may have gathered, I find Sanderlings rather compelling little birds to watch.

They're pretty impressive close up, too. Slightly easier to approach than some other shorebirds, good views of them will show that their stark white bodies contrast strongly with their dark bills and eyes, rather like a snowman with a couple of coals for peepers. Their running action is unlike any other bird, too – their legs move as if they're cycling. All in all, the Sanderling is a little cracker.

Where and when?

In rough weather, Sanderlings can be driven inland and can occur on a variety of types of wetland

MARTIN'S QUICK QUIZ

Once you've seen your Sanderling, can you tell the main difference between its feet and those of other small waders? Answer on page 251.

Common Sandpiper

Wood Sandpiper (p162)

Green Sandpiper (p161)

Common Sandpiper

Scientific name: *Actitis hypoleucos*

Approximate size: 20cm

Wingspan **2m**

1m

The word 'sandpiper' can be a bit tricky. It's used to describe some members of the stint family (see page 151), but it also refers to what you might call the 'true' sandpipers. There are three regular members of that family in the UK, and they're rather similar, so the photos opposite are your best starting point.

We'll begin with the Common Sandpiper, which actually lives up to its name, and is indeed the commonest of the three. It breeds on fast-flowing streams and rivers with stony shorelines and beaches, generally in upland areas.

Your best bet is to see them during the migratory seasons (they also breed in Scandinavia), but they do also winter in small numbers on estuary creeks, rivers, docks and reservoirs – frankly, anywhere that won't freeze up and has enough mud to work over in search of food.

Where and when?

MARTIN'S QUICK QUIZ

The Common Sandpiper's posture is rather crouched and leaning forward. Why do you think this is? Answer on page 251.

This is a bird that bobs, rather like a clockwork toy. Its rear end is almost always bobbing up and down, as it busily walks the length of a scrape and back again, making short flights to cross water or obstacles. In flight it flies low over the water, with bowed shivering wingbeats and head held up, in short glides. The obvious broad wing bar and white underwing can be seen, which the other two species covered lack.

Look for sandy brown upperparts with darker centres to the feathers and barring. You'll also notice a very long tail, protruding well beyond the wingtips. The key feature, however, is a very clean breast band with an obvious white crescent on the side.

All three of the sandpipers can be seen alongside each other at times during peak passage periods

Green Sandpiper

Scientific name: *Tringa ochropus*

Approximate size: 22cm

This bird spends the winter here in small numbers, often found in ditches, farm puddles and similar places. However, you often won't see it until you flush it, when it towers into the sky, calling, and flies away into the distance, looking like a large House Martin. Green Sandpipers roost together and will feed in loose groups in places where the mud is productive. Scrapes are a great place for these birds and counts of more than 30 can be seen at a single site.

In comparison with the other two sandpipers, look out for very dark upperparts with fine pale spots and white underparts separated on the breast by a clear-cut border. Check also for broad uppertail bars, greyish-green legs, a dark head with a broad white stripe in front of and above the eye, and a white eye-ring.

Wingspan 2m

1m

Where and when?

GWC score 4

Wood Sandpiper

Scientific name: *Tringa glareola*

Approximate size: 20cm

Wingspan 2m

1m

And so to the third of our sandpipers. This delicate wader breeds in very small numbers in northern Britain, but most commonly in Scandinavia and Russia on marshes and bogs in the Taiga belt. It migrates through in the spring or early summer in small numbers, and regularly occurs in autumn, especially after easterly winds.

The key features to look out for here are the browner upperparts with large pale spots, which create a sort of spangled look. The Wood Sandpiper also has a dark cap, a prominent whitish stripe above the eye and darker stripe through the eye. The light streaking on the breast does not end abruptly as it does on the Green Sandpiper.

In flight, the lighter underwings and toes that project beyond the tail help with identification but a quieter call than that of the Green Sandpiper is a major giveaway.

It is longer necked than the other two, and the longer yellowish legs make the bird more graceful in appearance, especially when it is actively feeding

Where and when?

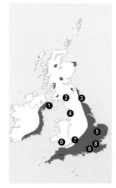

Ringed Plover

Little Ringed Plover (p167)

The Ringed Plover's bill has an orange base and dark tip, whereas the Little Ringed Plover's is darker and thinner

Ringed Plover

Scientific name: *Charadrius hiaticula*

Approximate size: 19cm

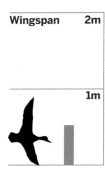

Wingspan 2m

1m

If you've been thumbing through this book page by page, you'll have noticed that we've already covered three species of plover – Lapwing, Golden and Green. So why have we put these two birds into a separate section? Well, the whole idea of this guide is to help you cut your first teeth as a birdwatcher. The first time you come across a Ringed Plover, you won't be turning to the 'plover' section of your guide. You'll want to turn to 'small wading birds', because to the first-timer, that's exactly what it is.

The first time I saw a Ringed Plover, I had no idea what it was, other than that it was about as cute a little scuttler as I'd ever come across. This tiny mouse-like bird, with its black bandit stripe and energetic pecking action, is like the Jack Russell of the wader world. It may be smaller than most, but no-one's told it so, and it mixes with the bigger birds as if it's got every right to.

Be careful when you're talking about it, though. Don't say 'look at that cute little Ringed Plover over there', because the Little Ringed Plover is actually another bird altogether. You can see the differences between the two in the photos opposite, but the main talking point is the variance in bulk. The Ringed Plover looks as if it has been pumped up, so that its head and tail are plumper.

Where and when?

The Little Ringed Plover, on the other hand, looks like the fitness freak of the two, with a slimmer build and longer legged appearance. Check out the heads, too. The Little Ringed has a browner cap, showing white between that cap and the band over its head, giving it the impression that it's wearing a helmet. It also has a yellow eye ring, which the Ringed Plover lacks. The Ringed has orange legs, while the Little Ringed's legs are pinkish, and the black bands around the neck and head are thicker on the Ringed.

You'll find the Ringed Plover on sandy shores and beaches, but it will also breed at gravel pits and scrapes, particularly if there's low vegetation. It's a very sociable bird, and can often be seen hanging out in Dunlin roosts.

Juvenile Little Ringed Plover

Juvenile Ringed Plover

Little Ringed Plover

Scientific name: *Charadrius dubius*

Approximate size: 14cm

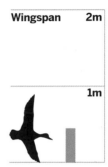

Wingspan **2m**

1m

Say hello to a fairly new colonist to Britain. The Little Ringed Plover – or LRP as Martin calls it – has taken full advantage of the gravel extraction across the country in recent decades, as the gravel pits that result provide perfect nesting habitats for a bird whose natural home is along with pebbly or shingle banks of rivers.

It's one of the first spring migrants to turn up, and you should look out for the twisting, turning display flight it performs. It's still an uncommon breeding bird, and its ground-nesting habits make its young vulnerable to Herons, Kestrels and others, but it has managed to gain a reasonable foothold over the years.

You can sometimes see juveniles feeding on scrapes with low water levels, and to help you distinguish them from Ringed Plover young, look for the more streamlined body, thinner bill and yellow eye-ring.

Where and when?

If there's a single small plover running around a freshwater scrape, it's most likely to be a Little Ringed Plover

Turnstone

Scientific name: *Arenaria interpres*

Approximate size: 23cm

Here's a very unmistakeable bird, partly because it doesn't really look like a wader. It's stocky and bull-necked, with short thick orange legs and a short pointed wedge shaped bill, which is great for flipping pebbles over, rather as you might have guessed from its name.

This high Arctic breeder migrates from Greenland and Siberia to our coasts where it is common on rocky, pebbly and sandy seaweed-strewn shores. It does occur on scrapes, estuaries or gravel pits during extreme weather such as fog, heavy rain or thunderstorms so it may turn up singly or in small flocks in spring or autumn.

Outside the breeding season, it looks rather dirty, as if someone has smudged its plumage. Good for camouflage, though. In breeding plumage, however, it transforms into a beautiful bird with bright orange back and bold black stripes.

In flight the Turnstone looks pied, with a large white lozenge on the back and a large amount of white on the tail

Wingspan   **2m**

1m

Where and when?

Great Waterbird Challenge
Part 4: What do you mean, 'just a seagull'?

After a fine lunch in the restaurant back in the main grounds of Slimbridge, we set off once more, pausing to check the hedgerows and bird feeders near the visitor centre. We were already fortified by warming food, and the House Sparrows, various finch species, and even a pair of late Swallows that flew overhead boosted our spirits even further. We were getting on towards the 70 mark, and it was looking good.

Martin pointed out the facility where staff at Slimbridge are raising **Cranes** (p183) for release in Somerset as part of the Great Crane Project. 'All going well, we'll end up with regular breeding colonies of Cranes in this country again,' he said. 'Britain's tallest birds will be back, once more.'

'Talking of tall birds,' I said, 'we haven't seen a **Heron** (p174) yet.' 'Good point,' he nodded. 'Come on, the next place should fill that gap.'

Before long, we were out towards the back of the reserve, where flocks of gulls were wheeling and shouting. 'The reedbeds over to the right there are wonderful water filters,' Martin explained, 'so we've got several types of freshwater fish living here. Where you find fish, you find Herons.'

And we did. Within seconds. We also spotted a **Little Egret** (p180) probing around in the distance.

'Fancy trying a spot of gull identification?' Martin asked, looking up at the skies. I peered upwards and squinted against the autumn sun. 'Um,' I said articulately. 'This looks tricky.' He smiled. 'Come on, back to the estuary.'

Back on the open flats, the flocks of gulls were settled on the mud, and were so much easier to see. Martin is a big fan of gulls, because as he points out, they come in so many

forms and there's so much to discover about them. Deep down, I was forming a rather different opinion. When a bird can have up to nine different plumage phases in the first few years of its life, what chance do you have of being able to sort them all out? I'll be honest, gulls don't really grip my imagination.

Thirty minutes later, however, I was beginning to change my mind. I could already work out which were the **Great Black-backed Gulls** (p204), huge bruisers that tower over all the others and stand around like bouncers at a gull nightclub. I was noticing that **Black-headed Gulls** (p192), even without the chocolatey-coloured heads that make them so easy to identify during the summer, had more of an angular appearance than the other gulls as they floated around on the water. Thanks to the arrival of a **Yellow-legged Gull** (p201), I discovered how to tell the difference between them and **Herring Gulls** (p200). The lesson here was clear. Gulls are indeed difficult birds for the newcomer to birdwatching, but if you're prepared to give them a go, there's actually plenty to discover. And if you can do so in the company of an enthusiast like Martin, then all the better.

As a **Cormorant** (p186) flew by, we pulled ourselves away from the gullish soap opera that had been unfolding before us, and made our way back to the reedbeds. Martin saw a Stock Dove, I saw a Wren, we both saw some Rooks and Jackdaws, and our total was well into the 80s. As we reached the reedbed, Martin froze. What looked like a large bat was flying to and fro across the water – back and forth, back and forth, as if it had lost something and was retracing its flight in an attempt to remember where it was. It was a **Black Tern** (p208). 'That's a real stroke of luck,' whispered Martin. 'One of those birds, like the phalarope, that just suddenly appears.' We watched the delicate creature for some time. Our total was now 84 – just 16 to go – but the star tern before us had gripped us. Just a few more minutes watching. Just a few more…

Now turn the page to find out more about the herons, gulls and terns – the tall birds and the noisy white birds.

TALL BIRDS

The birds we've covered so far have been, I suppose you could say, 'L-shaped'. When you look at them, mainly on the water, their bodies are essentially horizontal, while their necks are relatively vertical.

This next set of birds tends to stand more upright. They really are genuinely tall, too, and include the tallest of our birds, wetland or otherwise. You'll find various members of the heron family in the pages ahead, along with a few rather special extras.

Grey Heron — Patient fisher with grey body and perky plume (p174)

Bittern — Brown-streaked with camouflaging black markings. Very shy (p177)

Cormorant — Often sits with wings spread, glossy plumage, hooked bill (p186)

Spoonbill — All white, but just look at that bill. Two huge ladles (p184)

Crane — Unmistakably huge, with red patch on head, and large plumed bustle (p183)

Little Egret — Twitchy and busy, all white with dark bill and legs, and yellow feet (p180)

Grey Heron

Scientific name: *Ardea cinerea*

Approximate size: 94cm

Let's start this section of tall birds with probably the most familiar one of them all. If you see it moving, then it's most probably in flight, because herons are notorious for their ability to stand absolutely motionless on the edge of a lake or river, waiting for precisely the right moment to strike at prey.

When it's resting, and doesn't need to lunge after a fish, a heron will stand with its long neck folded deep into its body, appearing hunched and thoroughly fed up. It might well stand on one leg. But when hunting prey, it holds its neck in a snake-like shape, twisting it and stretching, but always keeping its eyes still and focused until it strikes. And the strike is something to see! It involves completely plunging into the water to spear or grab anything from rodents to amphibians or ducklings, and particularly fish, with its strong dagger-like bill.

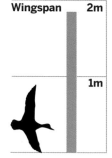

Wingspan — 2m / 1m

Where and when?

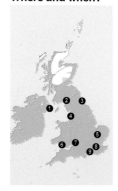

MARTIN'S QUICK QUIZ

Two other heron species seen sometimes in Britain are also named after their colours. What are they? Answer on page 251.

The juvenile lacks the long breast feathers and black wing edge

If there's a strong wind, herons will tend to stand in a favoured sheltered spot

A large bird with long, broad rounded wings in flight, it appears from a distance grey-backed with a black edge to the wing and a striking white neck. On the head, a broad black stripe from behind the eye carries on to form an 'aerial' that sticks out behind the head, very fitting for a bird that often has deliberate robotic 'remote control' movements.

Herons nest quite early in Britain, from about February onward, and the young, which appear from late June, are somewhat darker and plainer than the adults.

Bittern

Scientific name: *Botaurus stellaris*

Approximate size: 75cm

Many birdwatchers have spent years hoping for a glimpse of a Bittern, but without any joy. Yet I also know of at least one person who has seen one on their very first visit to a reedbed. That's the thing about Bitterns, you see – they don't appear very often, but when they do, they're unmissable.

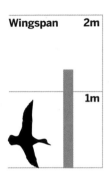

Wingspan   2m

1m

Where and when?

You're far more likely to hear one than to see one, though. During the months of spring, male Bitterns, like other birds, are prone to vocal activity.

Unlike other birds, however, the Bittern has a voice like a foghorn, 'booming' its call by sucking in air that inflates the throat. Hear it once and you'll never forget it, I promise.

Should you want to see what you hear, then look out for a large, stocky brown and buff heron with a rather hunched appearance. It relies on incredibly well-camouflaged plumage to hide in its favoured habitat of reedbeds, through which it moves stealthily while waiting patiently for a meal, which is usually fish or frogs.

During cold weather, it is more likely to be seen in the open, catching the winter rays of sun, or walking on ice in search of food. Martin has even seen them catching voles that were underground at Slimbridge!

In flight it can look like a giant owl or bird of prey and is often mobbed (chased and harassed) by other birds which see it as a potential threat.

If it is surprised, the Bittern stands upright, stretching its neck, and will actually sway with the movement of the reeds, possibly the finest master of camouflage in the British bird world

MARTIN'S QUICK QUIZ

The scientific name of the Bittern – *Botaurus* – relates to the bull-like nature of the bird's bellow. True or false? Answer on page 251.

Little Egret

Scientific name: *Egretta garzetta*

Approximate size: 60cm

Wingspan   2m

1m

This bird is a slim and graceful medium-sized white heron that has rapidly colonised the UK since the late 1980s. It can be found nesting in colonies, and is quite often spotted in existing heronries among Grey Herons.

The Little Egret is not a hard bird to recognise, thanks to being completely white with long, black legs and a long, very sharp bill. It's more of a pleasure to watch than the Heron, as it tends to be much more active in its hunt for fish and amphibians. In fact, it doesn't just walk around looking for prey, but even runs about with wings spread open.

So how does that work? Well, if you look closely, you'll see that Little Egrets have bright yellow feet. Martin tells me that the movement of these bright appendages can disturb, confuse and even attract fish towards it. A little tip for anglers, there.

Where and when?

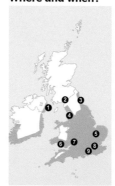

MARTIN'S QUICK QUIZ

Back in the 19th century, egrets were hunted for a certain type of fashion. What item was it? Answer on page 251.

Little Egret

The feet and skin between the eyes change to a reddish colour for a short period during the breeding season

You can find the Little Egret in various types of wetlands, from ditches to the open estuary. One good tip is to look at pools that appear to be drying up, as small flocks sometimes gather at such sites where the food is more concentrated.

There are two other types of egret I'd like to mention here: the Cattle Egret is increasing in the UK and has now actually bred, while a similar story is unfolding for the Great White Egret, which should also begin nesting here in future. For the present, however, if you see more than one egret in one place, it's almost definitely a Little Egret.

Common Crane

Scientific name: *Grus grus*

Approximate size: 115cm

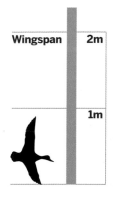

Wingspan | 2m

1m

This is a bonus bird, and it is rather special. Cranes have been largely absent from Britain for centuries, ever since they were hunted into oblivion, but they're making a tentative return. A few have appeared in Norfolk in recent years and, thanks to a programme by WWT, RSPB, Pensthorpe and Viridor Environmental Credits Company, a colony has been released in Somerset.

Once extremely common (if you live in or near a town beginning with 'Cran' or 'Tran', then cranes were once abundant there), it would be wonderful to see these most exciting of birds enjoying our wetlands once more.

KATE'S TOP TIPS

You want tall? This bird is tall – standing at around 1.2m in height

Where and when?

Spoonbill

Scientific name: *Platalea leucorodia*

Approximate size: 85cm

Wingspan 2m

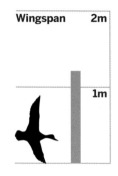

1m

Many views of this bird will be of a rather erect stance, with its bill tucked away in its wings, asleep. But when the Spoonbill begins to feed, then the fun really begins! It wades through the water, sometimes belly deep, while sweeping its head side to side until it makes contact with its prey. It will then stop to swallow the prey and then continue to work over the patch. After a little while feeding like this, it will suddenly fly off to try its luck elsewhere.

So there it is, a big white bird, bigger than the Little Egret. Anything else to distinguish it? I'm glad you asked. The Spoonbill isn't so-called just for fun. If you can get a good look at this amazing bird, you'll see what appears to be a pair of ladles sticking out of the front of its face. These spatulate bills help with the feeding style I mentioned above. Someone once said that they looked rather like a pair of tongs. You could even imagine the Spoonbill hanging its bill by the fireplace when it goes to bed at night.

Where and when?

This real treat of a sighting is a scarce summer visitor, although a handful can be found in the South West of England during the winter, Slimbridge being one reserve where they may turn up.

If you see a large white bird in flight, and aren't sure whether it's a Spoonbill or an egret, then check the neck. A Spoonbill flies with neck more outstretched than an egret

Cormorant

Scientific name: *Phalacrocorax carbo*

Approximate size: 90cm

The first time you come across a Cormorant roost, it can take you aback. As you gaze at the great dark birds, scattered among the branches of a tree on the edge of a gravel pit or lake, with their wings held out like pterodactyls, and their heads raised high like raptors, your first thought is that you've stumbled upon some sort of prehistoric scene. It has quite a spooky effect.

Closer to, you'll notice that the Cormorant isn't all that black after all. Its plumage, when looked at through binoculars, bears rather a striking blue or green glossy sheen, rather like a starling when seen at closer quarters.

These excellent fisher-birds can look a little like black geese in flight, but watch them carefully, and if they glide from time to time, then they're Cormorants. We have two races in this country, one which breeds near cliffs and rocks, the other in trees or reedbeds.

Wingspan **2m**

1m

Where and when?

MARTIN'S QUICK QUIZ

If you see a 'Cormorant' with a quiff, what type of bird are you actually looking at?

Answer on page 251.

Cormorants can often be seen in the middle of lakes or gravel pits, on posts, buoys and rafts built for nesting terns

NOISY WHITE BIRDS

I reckon most of us know what a gull looks like, but telling them apart is one of the tougher tasks for the novice birdwatcher. In their adult plumage, they're usually fairly straightforward, but they tend to take a few years to get there. A close look around a wetland should reveal a bunch of 'seagulls', quite likely five or more species, but there could be as many as four different plumages from each, depending upon the age of the birds.

Black-headed Gull — Common, with a dark brown head and dark red bill (p192)

The smaller gulls take just two years to mature into adults, but have five plumages during that period. The large gulls take four years to mature and have nine different plumages during this period.

All of these plumages overlap at one time or another, which can be confusing – but the more you get into gull-watching, the more fascinating it becomes. There is still a lot to be learned about them and you could help fill in the gaps. ▶

Common Gull – Has a delicate porcelain look, with yellow bill in summer (p195)

Little Gull – Very small and coastal, with a black head (p198)

NOISY WHITE BIRDS

For the purposes of this book, however, we will concentrate on the adults. Learn to recognise them, and you've actually made the first step towards knowing your juveniles, too. It's quite possible to be able to identify a younger bird based upon its size, shape and attitude – thanks to the knowledge you've built up about its parents. Get to know the six species in the pages ahead, and you're on your way.

Herring Gull — Red spot on yellow bill, greyish upperparts (p200)

Lesser Black-backed
Gull – Dark upperparts,
yellow legs (p202)

Great Black-backed Gull
– Huge, black upperparts,
pink legs (p204)

Black-headed Gull

Scientific name: *Larus ridibundus*

Approximate size: 36cm

Wingspan   2m

1m

This small gull is probably the gull you're most likely to see, and in the summer months at least, very easy to identify. In breeding plumage they sport a lovely rich brown head, which looks as if it's been dipped in cocoa-rich chocolate. During this time, they're fascinating to watch, as parent birds in their busy colonies go about the business of raising their young.

Once the young have fledged – left the nesting area – they become amazing peach, white and grey coloured birds for a short period. This is their juvenile plumage, which quickly turns into the first winter garb. In winter, adults lose their chocolatey colouring, but can still easily be identified by a dark mark through and above the eye, and a dark spot just behind it.

Where and when?

MARTIN'S QUICK QUIZ

Another species of gull, with a darker black head, can sometimes be seen among the Black-headed Gulls. Do you know what it's called? Answer on page 251.

Black-headed Gull,
winter plumage

KATE'S TOP TIPS

When sitting on the water surface, Black-headed Gulls have a very upright, perky swimming style

During the winter months, Black-headed Gulls can often be seen wandering around among flocks of waders in fields, not for the company, but to steal earthworms from them.

If there aren't any waders around, then look out for a nearby tractor plough or rotavator – the Black-headed Gulls will follow these just as they would the incoming or receding tides on the estuary.

Common Gull

Scientific name: *Larus canus*

Approximate size: 41cm

This fellow is larger, with a slightly darker back and wings than the Black-headed Gull, and is often found in the same company. This may be when it is resting or washing on floodplains, or walking around looking for invertebrates on fields and turf.

A lot of people often think of gulls as tough-looking bruisers, but not the Common Gull: it has rather a cherubic look compared to other members of its family. The head is white in summer, but streaked in winter, with the bill changing from a greenish-yellow in summer to a duller tone with a band around it near the tip. The eye is black. The legs are also greenish-yellow.

Winter's the best time to see Common Gulls as they gather in large roosts. During these months you can see long straggling lines or V-formations making their way to roosting sites on open water, or most typically, following tractor ploughs or on playing fields.

In adult plumages the large white 'mirrors' (as Martin calls them) on the wingtips are easy to see. In juveniles the scalloped, brownish plumage changes in its first winter.

Wingspan   2m

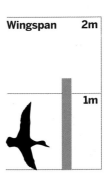

Where and when?

Although Common Gull is the traditional name for this bird, it's not our commonest species, and its name may come from its habit of feeding on commons

GWC score 4

Little Gull

Scientific name: *Larus minutus*

Approximate size: 26cm

Wingspan **2m**

1m

The last of our smaller gulls, this tiny and dainty member of the family is a full third smaller than the Black-headed Gull. In fact, Little Gulls are more reminiscent of terns than the other gulls due to their behaviour while feeding. They feed mostly in flight, picking items from the surface of the water with their bouncy flight and rounded wings.

The key feature to look out for is the full black hood it sports in breeding plumage, although its much smaller size helps to differentiate it from the Black-headed Gulls. As it flies past, see if you can get a glimpse of the underwings – they're quite a dusky colour. The wingtips are white.

You can find Little Gulls on freshwater sites on migration or after gales, most likely from late March to May or from August to October, but usually in small numbers.

Where and when?

Young Little Gulls, up to a year old, will have a dark 'W' across the upper part of the wings

*Little Gull,
first winter*

Herring Gull

Scientific name: *Larus argentatus*

Approximate size: 60cm

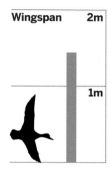

Wingspan　**2m**

1m

You'll know this bird by the call alone. It's the classic seaside sound, and if you think of British beaches, buckets and spades and deckchairs that don't work, then the gull call which accompanies that image in your imagination will be that of the Herring Gull. There's no such single species as a seagull, but the Herring Gull is the closest to claiming that name.

Herring Gulls are large and heavily built, with something of a mean look in their eye, as if they reckon they rule the roost and have every right to. They've got pink legs, and a stout yellow bill with a red spot on the lower part. That red spot is a good identification point. The steely eye is yellow with a yellowish ring around it.

Of all the large gulls we encounter regularly, the Herring Gull has the palest grey upperparts when in adult plumage, and this shows in second-year birds onwards. Although they're seabirds, Herring Gulls have increasingly come inland over the years, and can often be seen resting or feeding in fields, or loafing around on freshwater sites. They've become quite common in some cities now, too.

Another fascinating change in Herring Gulls in recent years is that they're not all, in fact, Herring Gulls. Martin pointed one bird out

Where and when?

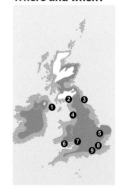

Yellow-legged Gull

KATE'S TOP TIPS

The amount of grey on the Herring Gull's back increases with age, just as with humans!

to me in a flock, and I noticed it had yellow legs, not pink ones. 'Some years ago,' he said, 'the yellow-legged birds were seen as a subspecies of the Herring Gull. They're now a species in their own right, and they're called, guess what? Yellow-legged Gulls.' Which is good news for the Great Waterbird Challenge. If I'd tried this a few years ago, the flock of Herring Gulls I saw would have given me one tick. As the Yellow-legs were split into their own species in 2007, I've got two!

Lesser Black-backed Gull

Scientific name: *Larus fuscus*
Approximate size: 58cm

Wingspan   2m

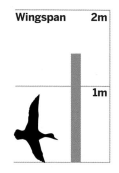

1m

The first of our two black-backed gulls is about the same size as a Herring Gull, although fractionally smaller. It's got darker plumage, though, and appears somewhat slimmer. The wings are longer, too – when it's standing, you can see the tips extend a tad further than the tail.

If you're looking at a black-backed gull all by itself, however, and you're not sure if it's a Lesser or a Great, then here's a good rule of thumb. Lesser Black-backed Gulls have yellow legs – and to help you remember that, 'lesser' has a double letter in the middle, as does 'yellow'.

You'll find this gull around the coast and lakes of Britain, although some colonies have moved into cities. Lesser Black-backed Gulls are big fans of rubbish tips, as well.

Large numbers winter here, although the species has declined in recent times, which is something of a worry as Britain is home to over a third of the entire world population which stands at some 300,000 pairs.

Where and when?

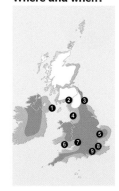

If you're near a reservoir, keep an eye out at dusk, as large numbers of Lesser Black-backed Gulls use these sites as night-time roosts

Great Black-backed Gull

Scientific name: *Larus marinus*

Approximate size: 71cm

Wingspan   2m

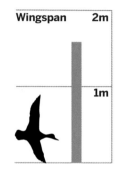

1m

The Great Black-backed Gull really is the head honcho of the family. The big cheese. Herring Gulls might look large and fierce, but stand one next to a Great, and it will suddenly have a decidedly wimpish look.

These huge beasts really can't be confused with any other gulls due to their sheer size – in fact, they're even larger than many geese. If you see a flock of gulls, with one with black wings standing head and shoulders above the rest, it's a Great Black-back (or a Lesser Black-back standing on something!)

Great Black-backs have large, deep, thick bills, and look like real bruisers. The upperparts are nearly completely black (much darker than in the Lessers), and the white spots, or 'mirrors', on the wingtips are huge. If you're still uncertain which black-back you're looking at, then where the Lesser has much yellower legs.

In flight, Great Black-backs reveal their very long wings, which result in a more laboured flight than other large gulls.

Where and when?

KATE'S TOP TIPS

Young birds have very dark bills which help mark them out from other young gulls

Common Tern

Scientific name: *Sterna hirundo*

Approximate size: 33cm

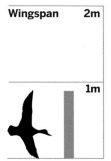

Wingspan 2m

1m

This is the tern you're most likely to see, as it's found throughout the summer months breeding on islands and rafts and platforms specially built for them at reserves such as WWT's. It likes clear water to breed alongside, as it dives for its fish, after hovering above the surface and scanning. If you see a nest site, do watch it for a while – it's great to see the returning male, fish in beak, arrive and, after a ritualised display, pass its catch over to its mate.

You don't want to get too close, however. Terns are notoriously aggressive defenders of their colonies – a particular trait of ground-nesting birds – and you don't want to be in the middle of their horror-film screams and fearsome dive-bombing.

Their attitude, however, is good news for other birds, and once the terns have moved in, defending their territories from marauding gulls, then other birds, such as avocets, often move in to take advantage of the relative safety.

Sleek, with elegant black caps and longer bills than those of the gulls, Common Terns' more acrobatic flight makes them stand out from the other noisy white birds we've looked at so far.

Where and when?

The scientific name of the Common Tern, 'hirundo', refers to its swallow-like, or hirundine, flight style

MARTIN'S QUICK QUIZ

There are several other types of tern you might find at a wetland site. I've pictured four of them here – the Arctic Tern, Sandwich Tern, Little Tern and Black Tern – but which is which? Answer on page 251.

Great Waterbird Challenge
Part 5: Nature red in tooth and claw

To tell the truth, we spent far longer watching that Black Tern than was wise. It was wonderful, but not wise. The time was now approaching 5pm, and with dusk not too far away, we suddenly realised we'd better get our skates on. A distant Buzzard got us up to 85, and a few minutes scanning a reedbed brought us a **Reed Bunting** (p244) and the good fortune of a glimpse of a **Cetti's Warbler** (p242).

This is where Martin's knowledge of the site so comes in handy. He thought for a moment, then we whisked off towards a field where he was banking on Linnets and **Pied Wagtails** (p232). Both came up trumps, and we picked up a few other bonuses, too. 'Little birds, little birds,' Martin kept muttering to himself, trying to think of some of the common or garden species that we might have overlooked. Birds like Collared Dove and Meadow Pipit helped our cause, and we kept inching through the 90s.

As the evening gloom started to gather, a Jay brought us up to 99, and there we stopped. We were back near the centre by now, and Martin had a face like thunder. 'Seriously, Martin,' I tried to comfort him, 'we've done really well. And look, it would be astonishing to hit the 100 on my first time out. Besides, it means I've still got something to shoot for in the future.'

But I confess, I was disappointed, too. We sat on a bench and watched a huge flock of Starlings swirling like a single organism in the fading light, gathering for their roost. It was at least a wonderful sight to finish with. The birds, Martin reckoned, must have been numbering some tens of thousands, and as the cloud that they formed billowed and swayed, they instinctively flew as one, each individual aware of the immediate proximity of its neighbours just centimetres away, and not a single collision as a result.

Then something remarkable happened. As one, Martin and I jumped up from the bench. As one, we turned to each other. As one, we gasped a single word. 'Starling!'

Frantically checking our notebooks, we scanned through the lists. We'd seen Starlings throughout the day, and yet neither of us had really noticed them. Such a common, well-known bird had hidden in plain sight all day long. We had our 100, and it was the dear old Starling that had brought us in.

Then something even more remarkable happened. As we turned with grateful faces towards the mighty flock that had been our salvation, it parted. It split into two, and like the cells of an organism, kept on splitting. Martin raised his binoculars. 'I don't believe it!' he gasped.

A **Peregrine Falcon** (p220) was trying its luck with the flock. Swooping at the massed ranks of Starlings, its speed and the sheer numbers of potential prey suggested it had every chance of catching a late supper for itself. The cloud split and re-formed as the falcon flew through, its dizzying numbers and continued close formation spoiling the raptor for choice, yet confusing it at the same time. After what seemed like an age, the Peregrine decided to call it a day, and disappeared over the distant treeline. The Starlings re-formed, and began settling in the trees.

Martin and I looked at each other. What a brilliant finish to a brilliant day. We'd made our target – in fact, we'd actually recorded 101 species in all, and I'd loved every minute of it. 'Fancy a pint?' I asked. 'Yup. Let's go celebrate.' It was getting quite dark, but I could see the relief on Martin's face. 'First round's on me,' I promised.

Then a Song Thrush hopped out of a nearby bush, clucked at us, flew off, and we had 102.

Martin laughed. 'See?' he said. 'It's easy.'

Now turn the page to find out more about small birds and hunting birds.

HUNTING BIRDS

Most bird of prey species – known as raptors – fly over British wetlands at one point or another on passage and some are resident and take full advantage of what is on offer. There are two species that can truly be considered wetland birds, and they are the Marsh Harrier and Osprey. A third species that is also linked to wetland birds is the Peregrine... as it has a habit of eating them! A fourth is the Hobby, which is agile enough to be able to catch dragonflies in flight.

Marsh Harrier – V-shaped wings, lengthy tail (p215)

Osprey – wings bent at the elbow, dark brown and white colouring (p218)

These four are the species we're covering in this book, as they have special wetland skills. But, just as you might see a Blackbird by a lake, even though it's not considered a wetland specialist, there are other birds of prey you might well see hunting over a wetland, too. We're not covering them here, but to help you with recognition, here are the four raptors you'll be able to read about, along with pictures of the more urban Red Kite, Buzzard, Kestrel and Sparrowhawk for comparison.

Red Kite — Distinctive forked tail and chestnut underbody

Buzzard — Flies with fanned tail, and rounded wings

Kestrel — Longish tail, and the familiar roadside ability to hover

Peregrine — Long, pointed wings, fairly short tail (p220)

Hobby — Very long, thin wings, darting flight like a large Swift (p222)

Sparrowhawk — A degree of orange on the under-belly, and barred tail

Marsh Harrier

Scientific name: *Circus aeruginosus*

Approximate size: 52cm

There are several birds that come to mind when you think of reedbed life – Bitterns, Bearded Tits and several warbler species among them. But there's only one that really sums up the space above those reeds, and that's the Marsh Harrier. Large and languorous, it drifts above its favourite environment, head turned downwards and wings held in a V shape, peering through the reed stalks for small birds and mammals that are too busy getting on with their own lives to notice the dark shadow looming above them.

Female Marsh Harriers are larger than males, with much darker bodies, with creamy yellow-white caps and bibs. The males have russet bellies, pale wings with black feather tips and streaked upper bodies and underwings. Juveniles resemble the females.

If you do see a Marsh Harrier dropping down into a reedbed, it's not necessarily zeroing in on its prey. The birds usually build their nests within the reeds, which I imagine would make a rather uncomfortable discovery for any Water Vole that might happen to wander that way! By the 1970s, the Marsh Harrier had almost become extinct as a breeding bird in this country, but thanks to the reduction of pesticide use and number of other conservation measures, we now have several hundred breeding pairs once more.

Wingspan   2m

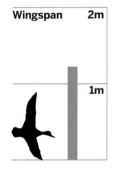

1m

Where and when?

MARTIN'S QUICK QUIZ

The Marsh Harrier is the biggest of the UK's three regular harriers, and the only one that is consistently seen over reedbeds. Do you know the names of the other two? Answer on page 251.

KATE'S TOP TIPS

In flight, look out for a long tail and a rather 'buoyant' glide of the Marsh Harrier (see page 212)

Female

Osprey

Scientific name: *Pandion haliaetus*

Approximate size: 56cm

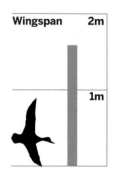

Wingspan   2m

1m

Quite possibly the greatest of all wetland sights, the Osprey has made the most welcome of returns in recent years. Having lost its breeding status in the UK, it's now returning in greater numbers, and nesting at a number of sites across Scotland, England and Wales.

If you get the chance to see it at one of its nest sites with public viewing facilities, such as WWT's Caerlaverock, RSPB's Loch Garten, or The Wildlife Trusts' Rutland Water, then I guarantee you'll never forget this bird. A rich, dark brown body, below which the white underparts stretch right up to the head, is topped off by a dark bandit's mask that runs through its eyes to the back of its neck. Slight plumes towards the back of the head give this large raptor a somewhat square-headed appearance.

Away from nest sites, you're most likely to see the Osprey in flight, perhaps on its way to or from Africa, where it overwinters. Seen from below, it's actually an unmistakeable bird – the underparts are predominantly white, with some mottling, while the wings are very long, and bend at the 'wrist', like no other bird of prey's do.

The Osprey is one of the UK's greatest recovery stories – long may it continue to build in numbers.

Where and when?

If you catch sight of an Osprey in flight with a fish that it's caught, you'll notice that it always holds the fish face-front to reduce wind resistance

Peregrine

Scientific name: *Falco peregrinus*

Approximate size: 42cm

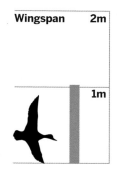

Wingspan 2m

1m

The Porsche of the bird world, built for speed and power, the Peregrine is a force of avian engineering. When you first see one, often drifting lazily towards an open scrape, your first impression is of sleek yet easy control. With apparent indifference, the bird appears over a distant treeline, seemingly uninterested in the flocks of waders down below.

Then they catch sight of him, his streamlined shape puts them into panic, and they take to the air, trying to rise above the falcon's flight path. Suddenly, the gears change. The Peregrine kicks up higher into the sky, twists and turns to get its best vantage point... then, like a blue bolt, plummets down into the whirling flocks below. The Peregrine's stoop is so terrifyingly fast that it's been clocked at over 200mph, and the plover that it's set its sights on has virtually no time to get out of the way. Martin tells me that the spiralled shape of the Peregrine's nostrils possibly help ensure that it doesn't pass out at such a speed.

The Peregrine will nest wherever it can find height, such as cliffs, leading to the occasional brood being raised on high towers and cathedral spires. At rest, it has a blue-grey back with a distinctive black moustache across its white face. Look out for the spotted breast, which helps distinguish it from other similar-sized raptors.

Where and when?

Adult

Juvenile

Hobby

Scientific name: *Falco subbuteo*

Approximate size: 33cm

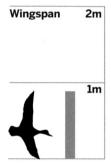

Wingspan   2m

1m

This small bird of prey has truly remarkable skills – it's one of the few creatures that can actually catch dragonflies in flight! If you perch yourself near a riverbank or a gravel pit on a bright sunny day, where large dragonflies such as hawkers zip and hover above your head, then just wait, you could be in for a real treat. Hobbies are so agile that they're able to match the acrobatics of a large dragonfly and pluck them in mid-air, whereupon they glide and feed on the morsels that they're clutching in their talons. As you watch, you can even see the dragonfly wings flutter to the ground as the Hobby consumes its meal before raising its head and banking off towards its next catch.

In flight, these rapid raptors, with their long, pointed wings and tilting, rapid flight, resemble Swifts – which is rather ironic, as they're even occasionally able to catch Swifts as well! The head has a grey cap and slight, dark moustache, rather like the Peregrine's, but not as pronounced. The body and underwings are heavily lined, but the key feature to look out for is the rich red colouring of the feathers that surround the top of the bird's yellow legs. It looks a little as if its own talons have bloodily wounded it near the base of the tail.

A summer visitor, Hobbies are best seen on warm bright days when hunting newly emerged, lethargic insects.

Where and when?

Although Hobbies gravitate towards woodland edges and heathland, high insect activity in late summer tends to bring them closer to wetlands

Cuckoo

Scientific name: *Cuculus canorus*

Approximate size: 33cm

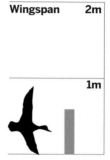

Wingspan 2m

1m

You might wonder why the Cuckoo has been placed in the Hunting birds section when it is, after all, not a bird of prey. Well, we had two reasons for including it here. The first is that the Cuckoo very strongly resembles raptors in flight, its long tail and pointed wings giving it a falcon-like appearance. As the purpose of this book is to help you differentiate between similar birds, no matter their family, then this placing should help.

The second reason was more to do with the bird's behaviour. Although they do not hunt smaller birds, they do seek out their nests, famously choosing the nests of Reed and Sedge Warblers, Reed Buntings and other little birds to lay their own egg in. Once a young Cuckoo hatches, it pushes its foster siblings out of the nest to their doom, leaving it alone to consume all the food its foster parents devotedly bring to it.

Cuckoos are often found in wetland areas – some of their favourite egg-host species are reed-dwellers as mentioned above – and their arrival in April and their distinctive call is celebrated as one of the great moments of the British springtime.

Where and when?

Young Cuckoo and Reed Warbler foster parent

Look out for strongly marked barring on the underside, with a blueish-grey back, rather similar to a Sparrowhawk. If you do see what you think might be a Cuckoo, try to watch it land. Cuckoos sit more horizontally than falcons and hawks, and tend to remain fairly conspicuously in the open, perhaps perched on a post, giving you the chance for a good look. And of course, should you hear its call – *cuck-oo, cuck-oo* – any lingering doubt will be removed.

SMALL BIRDS

No matter the type of wetland you're exploring, there'll always be a range of tiny tweeters, songsters, peepers and foragers, nipping in and out of the foliage and keeping your binoculars busy as you sort out one from the other. Many of them you'll be familiar with – Robins, Blackbirds, House Sparrows, Starlings, finches, tits and more are all reasonably common garden birds, and to delve into their various families is not the purpose of this book. If you want to brush up on garden and woodland birds, there are plenty of books to help you (see page 256).

Kingfisher – Bright, unmistakeable, and a must-see for any birdwatcher (p228)

Wagtails – Slim, horizontal stance and long vibrating tail (p232)

Reed Bunting – The sparrow of the reedbeds (p244)

But some of the small birds that you're likely to see around your local gravel pit or reedbed reserve are true wetland specialists. The Kingfisher, for example, may be the size of a sparrow, but you won't find it nesting in your ivy-clad wall. There are some warblers that will call ceaselessly from the silver birch at the foot of your garden, but there are others that will barely set foot outside of a reedbed other than to migrate.

So here, then, are the little chaps – the small birds that are the wetland equivalents of your garden favourites. And there are some real crackers among them.

Sand Martin – Aerial acrobat of the summer months (p246)

Wetland warblers – Easy to hear, tricky to see little brown jobs (p238)

Bearded Tit – Moustached rather than bearded, with long tail (p248)

Kingfisher

Scientific name: *Alcedo atthis*

Approximate size: 16cm

Well, we've put it into the small birds category, but to be honest, the Kingfisher could sit in a category all by itself. The jewel of the riverside, the splash of colour on a drab day, the flashing turquoise you glimpse from the corner of your eye: the Kingfisher commands attention wherever and whenever it's seen.

Surprisingly, however, it's seen much less frequently than you might think for such a shining gem. It's distinctive and unmistakeable, but its small size, shy behaviour and fast, direct flight can make it hard to see well. It's also something of a master of disguise. Head on, it appears orange and white with blue crown and wings, but from behind it is blue with an electric-blue streak down the back. What else to look out for? Well, it catches fish with its long dagger-like bill, which takes up almost a quarter of the bird's entire length.

Wingspan 2m

1m

Where and when?

MARTIN'S QUICK QUIZ

Male and female Kingfishers are virtually identical, except for one distinguishing feature. Do you know what it is? Answer on page 251.

The Kingfisher excavates a hole in a quiet waterside bank and its tunnel can be a metre long with a nest chamber. If you come across a Kingfisher bank during the breeding season (there's an excellent one at WWT's Slimbridge reserve, for example), then keep your distance, and you'll be able to see the adults fly to a regular perch, look around, then pop into the hole to provide their young with fish. It's an action they repeat every few minutes.

One of the great things about seeing a Kingfisher – other than the sheer visual sumptuousness – is the fact that they're great indicators of good, healthy freshwater systems. As they need clear water to feed by, their presence suggests high quality of water, and therefore a good ecosystem.

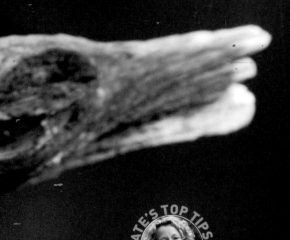

KATE'S TOP TIPS

Your first sign of a Kingfisher is usually a high-pitched 'chee'. If you hear that, look to the water's surface, as the bird flies fast and low just a few feet above the river

Pied Wagtail

Scientific name: *Motacilla alba*

Approximate size: 18cm

Here's a busy little bird that's probably already familiar from the car park at your local supermarket. In fact, they're so comfortable living alongside humans that it's quite possible you've got closer to a Pied Wagtail than any other wild bird. The boldest will even scuttle along pavements, nipping in between people's feet, as they seek out tiny insects scuffed up by pedestrian traffic.

They may have made their homes in urban centres, but Pied Wagtails are really waterbirds, catching flies and picking for windblown insects at the edges of ponds and lakes across the country. They don't hop, like most small birds, but walk or run, their tails wagging up and down furiously as they go. Even their flight 'wags' – it's a bounding action with deep bounces.

In Britain and Ireland we have our own version of this widespread bird, but throughout the rest of Europe a different form, called the White Wagtail, is found. This bird is similar, but can be readily identified, especially in the spring when it occurs as a passage migrant, as it has a greyer, lighter, back.

Globally, Pied Wagtails are commonly found only in Britain and Ireland

Wingspan 2m

1m

Where and when?

Grey Wagtail

Scientific name: *Motacilla cinerea*

Approximate size: 18cm

When you look at this picture, you might feel that yellow is actually the main colour that comes to mind, and fair enough. It certainly does sport bright lemon-yellow below its tail, and even more on the male during the breeding season. But take a glance at the Yellow Wagtail over the page, and you'll soon realise that that's the bird that's really got the dominant citrus colouring. The 'Grey' part of Grey Wagtail helps differentiate between the two, and in any plumage, from juvenile to adult, it has greyish upperparts and whitish underparts, with a whitish stripe over the eye and a black bib in summer (winter birds all have white throats).

If you're still uncertain, the season might help. Whereas Yellow Wagtails are summer visitors only, Greys can be found here throughout the winter months, and, indeed, are more likely to appear in urban areas during this time.

When breeding, however, they favour fast-flowing streams, rivers, mill pools and areas of water with stones to catch flies from. Like all wagtails, their rapid tail-bobbing helps identify them.

If you see a wagtail with a yellow belly on a rock in a fast-running stream, it's probably a Grey

Wingspan   2m

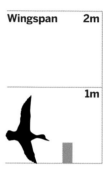

1m

Where and when?

Summer

Winter

Yellow Wagtail

Scientific name: *Motacilla flava*

Approximate size: 17cm

Wingspan 2m

1m

Some birds are yellowy, some are yellowish. This one, however, is the sheer epitome of yellow. Even from a distance, its bright-lemon underparts seem to shine at you. I reckon it looks as if you could tip it up, give it a squeeze and make some mighty fine pancakes. Not that you should ever squeeze any bird, of course. They really don't like it.

If you only get a glimpse of the top half of a Yellow Wagtail – perhaps it's half hiding behind a bank or tussock – then olive-green is the colour to look out for, in contrast to the black back of a Pied Wagtail or the grey of a Grey. In general, however, your best bet is actually to look up, as most Yellow Wagtails are seen during the migration periods, when they fly over at height during spring and autumn.

Where and when?

When they do settle, they regularly feed around wetlands favouring livestock, especially cattle and horses, which kick up the insects for them. Martin tells me that during our winter months they do exactly the same thing, but as they spend that time in Africa, it's elephants they favour.

As ground-nesting birds, they can often be seen perched quite conspicuously on fence posts or bushes.

Yellow Wagtails prefer to hunt for insects around marshy ground with ditches and pools

Female

Reed Warbler

Scientific name: *Acrocephalus scirpaceus*

Approximate size: 13cm

Wingspan   2m

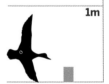

1m

If I tell you that the Reed Warbler is actually a reasonably common bird, you might wonder why it is that it scores two points, not one, on the GWC scale. Well, that's because Reed Warblers are far more often heard than seen. Perfect examples of the phrase 'Little Brown Job', they flit around in the middle of reedbeds, advertising their presence, but often offering less than a fleeting glimpse.

The Reed Warbler is warm brown above, paler below, and with a whitish throat and undertail. It is fairly slender with a long, pointed bill.

It sings a similar song to the Sedge Warbler (see over the page) but has a lot more stamina, singing for long periods without a break, but never getting too excited.

Where and when?

MARTIN'S QUICK QUIZ

Another type of warbler found in reedbeds gives off a trilling sound like a cricket. Can you guess what it's called? Answer on page 251.

KATE'S TOP TIPS

The bill and tail give the Reed Warbler a 'sharp at each end' appearance

Sedge Warbler

Scientific name: *Acrocephalus schoenobanus*

Approximate size: 13cm

Wingspan **2m**

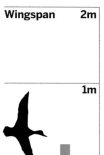

1m

This warbler spends its winter in Africa, and is typically with us from April to October. It favours reedbeds and ditches with emergent vegetation and taller sections of hedge, and bushes such as bramble, and trees such as willow to sing from. And this is the key thing about the Sedge Warbler. Unlike the other wetland warblers, it's quite happy to sit in full view and deliver its song, giving you plenty of time to study it.

Once it's settled into its territory, the males can be seen readily climbing up stems or branches to reach the top and sing their loud grating song. They're good mimics too, adding in the calls of other species to the mix. As the song reaches its high point, the bird will perch in the open and even make a song flight high above the perch before dropping back down into cover.

Where and when?

The Sedge Warbler is a streakily plumaged bird, but the most striking features are a thick whitish stripe above the eye and, when singing, a reddish inside to the bill. This is known as the gape.

The Sedge Warbler's song is performed in intense bursts, unless the weather is poor

Cetti's Warbler

Scientific name: *Cettia cetti*

Approximate size: 14cm

Wingspan 2m

1m

Cetti's Warblers are like the opposite of Victorian children – heard, but rarely seen. In fact, this is a bird that can really make you jump. It made me jump. There I was, peering over the top of a reedbed, wondering if there was something in the distance, when from just a few yards in front of me, a sudden blast of explosive song rocked my eardrums. 'That's a Cetti's,' whispered Martin. I pulled myself together, and stared intently at the spot where the ghettoblaster seemed to be. And that, my friends, was that.

The Cetti's is very much a skulker, preferring to remain in deep cover keeping an eye on you. If you do get a look at one (and hides provide the chance to do this) you will see a reddish-brown bird with a long rounded-looking tail, a greyish face and pale throat.

The Cetti's Warbler prefers reedbeds with scattered bushes that give dense cover. It is a fairly recent colonist from southern Europe, but, may suffer drops in the population in cold winters as it tends to remain here throughout the year.

The Cetti's Warbler is attracted towards the sounds of human activity, but can see you without you seeing it!

Where and when?

Reed Bunting

Scientific name: *Emberiza schoeniclus*

Approximate size: 16cm

Wingspan   2m

1m

This streaky bird looks very sparrow-like – and with good reason. It's a member of the bunting family, (which includes Yellowhammers, incidentally), which is closely related to sparrows.

You can see the Reed Bunting all year round, but it tends to move out of the reedbeds and ditches in the winter to find crops of seeds, or even sometimes to feed under or on bird tables.

During the breeding season, however, it uses any wetland with emergent vegetation, especially reedbeds, with scattered low trees and bushes. It is from these that it delivers its rather lonely song. Males are distinctive and easily seen when singing: they sit on an exposed perch and the combination of black head, bill and throat with a striking white moustache and collar really gives them away.

The plumage is otherwise like the female and immature birds', with rufous brown and buff-coloured streaks. It is a skulker but will feed on seeds on the top of vegetation and on the ground.

In winter, the plumage around the head wears away to reveal a rather ghostly version

Where and when?

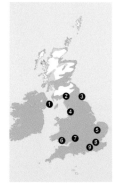

Sand Martin

Scientific name: *Riparia riparia*

Approximate size: 12cm

Wingspan 2m

1m

Although there's every chance that you'll see House Martins around wetland areas, it's the Sand Martin that is the true wetland specialist of the two. The great thing about martins is that as they twist and turn so much in flight, they give you a chance to see both the upper and the undersides of their bodies, which helps in identification.

First, let's look at them from above. Where Sand Martins have dark brown upperparts, House Martins sport a more spangly blue plumage and, importantly, have a white rump. When seen from below, the House Martin has completely white underparts, while the Sand Martin sports a brownish collar. Just think of the Sand Martin as being the more smartly dressed, as it's remembered to put on collar and tie, and keep its rear end properly covered up!

You'll find these aerial acrobats along rivers, gravel pits, lakes and any other water bodies where flying insects make cracking snacks.

Where and when?

You're unlikely to see a Sand Martin resting on the ground, but look out for them taking a breather on telephone wires, fences or branches

MARTIN'S QUICK QUIZ

You'll notice from the top of the opposite page that Sand and House Martins are known as hirundines. What is the other main member of this family in the UK? Answer on page 251.

Bearded Tit

Scientific name: *Panurus biarmicus*

Approximate size: 12cm

We'll finish our small bird section with a rather smart fellow, the Bearded Tit. Personally, I think it's a bit of a misnomer, as the dark feathers on the male's face resemble a moustache rather more than a beard. Think of a Mexican character from an old Western, and you've got the look of the Bearded Tit.

But there's more to this bird than just the 'tache. A subtle grey head, and a pleasant, soft fawn body are the other key features of the male, along with a long tail that wafts like a streamer behind the bird as it flies over reedbeds. Look out for whirring wings, too. Females are drabber, but the tail helps in identification.

You're only really going to see them in reedbeds, and the thing to listen out for is their noisy call – a 'pinging' sound that grabs the attention, and is usually the first indication of their presence.

Wingspan   2m

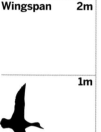

1m

Where and when?

MARTIN'S QUICK QUIZ

Are Bearded Tits members of the tit family? Answer on page 251.

'Beardies' are sociable
birds, and you'll rarely
see one by itself

P19 The male (cob), has a larger knob on its bill than the female (pen).

P36 It was once thought that Barnacle Geese hatched from Goose Barnacles

P42 Shelducks do indeed nest in holes or rabbit burrows

P50 A Teal with a vertical white stripe near its shoulder is a Green-winged Teal

P54 A Wigeon with Teal-like colouring on its head is, in fact, an American Wigeon

P57 Shovelers tend to be found in shallow water

P72 A Pochard with a bright red bill will probably be a Red-crested Pochard

P80 The Red-breasted Merganser tends to be more of a coastal or estuarine bird than the Goosander

P96 The Little Grebe is also known as the Dabchick

P114 The Curlew's remarkable bill allows it to dig deep into the mud for hard-to-reach lugworms

P119 Avocet nests are usually shallow scrapes on bare mud

P123 The stripe above a bird's eye is called the supercilium

P133 The Greenshank's bill turns very slightly upwards

P134 The Lapwing is also called the Peewit, after its call

P137 Golden Plovers fly high in an attempt to avoid Peregrines

P142 The other type of snipe found in Britain is the Jack Snipe

P147 Juvenile Knot are similar to the adults but with more of a washed-out peach colour on their breasts

P153 Temminck's Stint has yellower legs than the Little Stint

P154 The three waders are Oystercatcher, Redshank and Turnstone. The gulls are Black-headed Gulls.

P157 Sanderlings have forward-pointing toes, but no back toe

P159 The Common Sandpiper's posture helps it locate its food

P174 Green and Purple Herons are also sometimes found in Britain

P178 True. *Botaurus* does indeed refer to the bull-like nature of the Bittern

P180 In Victorian times, egrets were hunted for their feathers which were used to decorate fashionable hats

P186 A Cormorant with a 'quiff' is, in fact, a Shag, its more coastal cousin

P192 The gull with the blacker hood is the Mediterranean Gull

P208 A: Black Tern; B: Sandwich Tern; C: Little Tern; D: Arctic Tern

P216 The other two main harriers found in Britain are the Hen Harrier and Montagu's Harrier

P229 The female Kingfisher's lower mandible is partly orange-red rather than all-black

P238 The trilling warbler you might hear is the Grasshopper Warbler

P247 In addition to the martins, the Swallow is the other member of the British hirundine family

P248 Bearded Tits, also known as Bearded Reedlings, are in a family all by themselves

Checklist

USE THIS LIST TO KEEP A TALLY OF YOUR GWC SCORES EACH TIME YOU SPEND A DAY BIRDWATCHING

	Page	GWC score	Date	Date	Date	Date	Date	Date	Date
Avocet	118	4							
Barnacle Goose	36	3							
Bar-tailed Godwit	123	4							
Bean Goose	32	5							
Bearded Tit	248	4							
Bewick's Swan	22	3							
Bittern	177	5							
Black-headed Gull	192	1							
Black-tailed Godwit	121	3							
Brent Goose	38	3							
Canada Goose	34	1							
Cetti's Warbler	242	4							
Common Crane	183	5							
Common Gull	195	2							
Common Sandpiper	159	3							
Common Tern	206	2							
Coot	102	1							
Cormorant	186	1							
Cuckoo	224	4							
Curlew	113	2							
Curlew Sandpiper	151	4							
Dunlin	149	2							
Gadwall	58	3							
Garganey	62	4							
Goldeneye	73	3							
Golden Plover	137	3							
Goosander	78	4							
Great Black-backed Gull	204	2							
Great Crested Grebe	92	1							
Green Sandpiper	161	3							
Greenshank	132	3							
Grey Heron	174	1							
Greylag Goose	26	1							
Grey Plover	139	3							
Grey Wagtail	234	3							
Herring Gull	200	1							
Hobby	222	4							
Kingfisher	229	3							
Knot	146	4							
Lapwing	134	2							
Lesser Black-backed Gull	202	2							
Little Egret	180	3							

	Page	GWC score	Date	Date	Date	Date	Date	Date	Date
Little Grebe	95	2							
Little Gull	198	4							
Little Ringed Plover	167	3							
Little Stint	152	4							
Mallard	46	1							
Marsh Harrier	215	4							
Moorhen	100	1							
Mute Swan	16	1							
Osprey	218	5							
Oystercatcher	116	2							
Peregrine	220	4							
Pied Wagtail	232	1							
Pink-footed Goose	28	3							
Pintail	60	3							
Pochard	70	2							
Red-breasted Merganser	81	4							
Redshank	129	2							
Reed Bunting	244	2							
Reed Warbler	238	2							
Ringed Plover	165	2							
Ruff	126	3							
Sanderling	156	4							
Sand Martin	246	2							
Sedge Warbler	240	2							
Shelduck	42	1							
Shoveler	55	2							
Smew	83	5							
Snipe	140	3							
Spoonbill	184	5							
Spotted Redshank	131	4							
Teal	49	2							
Tufted Duck	68	1							
Turnstone	168	4							
Water Rail	104	5							
Whimbrel	115	4							
White-fronted Goose	30	3							
Whooper Swan	20	3							
Wigeon	52	2							
Woodcock	143	5							
Wood Sandpiper	162	4							
Yellow Wagtail	236	4							
TOTAL									

About WWT

WWT is a leading wetland conservation organisation saving wetlands for wildlife and people across the world. It was founded in 1946 by renowned naturalist and artist, the late Sir Peter Scott.

WWT's primary objective is to save wetlands and their wildlife by identifying and acting to counter the threats that affect their survival and, crucially, to enhance people's lives through learning about and being close to nature and inspiring them to take action.

From its humble beginnings at Slimbridge (pictured) in Gloucestershire, WWT now has a network of nine wetland visitor centres across the UK (see page 11), including nationally and internationally important wetland habitats totalling over 2,600 hectares. Our centres welcome more than a million visitors each year, and over 200,000 members add their support to our cause.

So why are wetlands so important?

Found from the poles to the tropics, from mountains down to the sea, wetlands are very diverse habitats, including lakes, ponds, rivers and their floodplains, marshes, swamps, coastal waters and, of course, canals. Yet half the world's inland wetlands have been lost over the past century and with them their unique wildlife. Wetlands are being lost and degraded more rapidly than any other ecosystem.

Wetlands are essential for life on Earth. As well as storing and cleaning our water, they can help protect us from floods and storms. They also include some of the most productive and diverse living systems – they are the lifeblood of our planet.

Wetlands provide habitats for a wealth of animals and plants, from flamingos to swans, from Marsh Marigolds to mangroves, from Water Voles to dragonflies.

Millions of people depend directly upon wetlands for their livelihood and, in an increasingly urbanised and frenzied world, many millions more enjoy wetlands as places to walk, relax and get closer to nature.

Wetlands also play an important role in the regulation of greenhouse gases. They store large amounts of carbon, and can help protect against the effects of climate change by reducing flood risk, stabilising shorelines and controlling erosion. Climate change is a major threat likely to have a significant impact on freshwater systems, with wide-ranging consequences for people and ecosystems. Unfortunately, wetlands remain highly threatened. Reclaimed for building or agriculture, increasingly polluted and degraded, wetlands and their wildlife are among the first casualties of our drive for growth and development.

The wildlife that wetlands support is among the most threatened of all ecosystems: of inland wetland-dependent species, one third of all amphibians, 15 per cent of wetland birds, over 40 per cent of reptiles, 30 per cent of mammals and 6 per cent of fish species are globally threatened with extinction.

Along with wildlife, the many benefits that wetlands provide to people are also at risk.

Visit a WWT centre Choose from WWT Caerlaverock in Scotland, WWT Castle Espie in Northern Ireland, WWT Llanelli in Wales, and our six WWT centres (Washington, Martin Mere, Welney, Arundel, London and Slimbridge) across England. Telephone 01453 891900 for opening times and admission prices, or visit www.wwt.org.uk/visit for details.

Become a member Members can visit all nine WWT centres across the UK as many times as they like all year for free, plus they receive a quarterly magazine, *Waterlife*. Visit www.wwt.org.uk/membership and sign up online or at your nearest WWT centre.

Adopt a bird Choose from a Mallard, swan, Néné, Eurasian Crane or flamingo and adoption packs are just £27 for a whole year. Visit www.wwt.org.uk/adopt.

Make a donation Support WWT's wetland conservation work in the UK and all over the world by making a donation either at your nearest WWT centre or at www.wwt.org.uk.

Want to know more?

FURTHER TITLES IN THE A&C BLACK WILDLIFE STARTERS' RANGE

Visit www.acblack.com/naturalhistory

BIRDS IN YOUR GARDEN This is the ideal companion for learning about the birds in your garden, featuring detailed profiles of all the species you are likely to see, and a wealth of useful advice about attracting birds to your garden and what to feed them. This practical and informative book contains all you need to know and is illustrated with many superb colour photographs and full-page portraits. It includes an introduction by Bill Oddie.

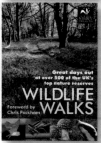

WILDLIFE WALKS Published in conjunction with The Wildlife Trusts, this superb guide comprehensively covers more than 500 of the UK's top nature reserves, all of them owned and managed by the unique network of 47 Wildlife Trusts. Each entry includes information on access, conditions, opening times, facilities, how to get there, and local attractions. It is illustrated throughout with many colour photographs and maps. *Wildlife Walks* is the only guide you'll need to plan a great family day out.